IX Brazilian Meeting on Simulational Physics (BMSP 2017)

Journal of Physics: Conference Series Volume 1012

Natal, Brazil
21 - 25 August 2017

Editors:

Bismarck Vaz da Costa
Paulo Henrique Martins
Lucas Alvares Mol
Luiz Felipe C. Pereira
Joao Antonio Plascak

ISBN: 978-1-5108-6289-0
ISSN: 1742-6588

Printed from e-media with permission by:

Curran Associates, Inc.
57 Morehouse Lane
Red Hook, NY 12571

Some format issues inherent in the e-media version may also appear in this print version.

Printed by Curran Associates, Inc. (2018)

For permission requests, please contact the Institute of Physics
at the address below.

Institute of Physics
Dirac House, Temple Back
Bristol BS1 6BE UK

Phone: 44 1 17 929 7481
Fax: 44 1 17 920 0979

techtracking@iop.org

Additional copies of this publication are available from:

Curran Associates, Inc.
57 Morehouse Lane
Red Hook, NY 12571 USA
Phone: 845-758-0400
Fax: 845-758-2633
Email: curran@proceedings.com
Web: www.proceedings.com

Table of contents

Volume 1012

IX Brazilian Meeting on Simulational Physics (BMSP 2017)

21–25 August 2017, Natal, Brazil

Accepted papers received: 17 April 2018
Published online: 3 May 2018

Preface

Papers

This special issue, *Celebrating the 20 Years of the Brazilian Meeting on Simulational Physics* (BMSP), includes invited and selected articles of the *IX* (BMSP), held in Natal, Rio Grande do Norte, Brazil, from 21st to 25th August, 2017. This is the ninth such meeting, and the third one to have contributed papers published in the *Journal of Physics: Conference Series* (the other ones were the VII and the VIII BMSP).

The previous meetings in the BMSP series took place in the mountains of Minas Gerais, in the region of the Brazilian Pantanal, and in the shores of Paraíba and Florianópolis. Now, for the first time, the Meeting was held in Natal, with its pleasing shores, the capital of the Rio Grande do Norte state.

The IX BMSP brought together about 50 researchers from all over the world for a vibrant and productive period. As in the previous meetings, the talks and posters highlighted recent advances in applications, algorithms, and implementation of computer simulation methods for the study of condensed matter, materials, out of equilibrium, quantum and biologically motivated systems. For the first time the meeting held two excellent tutorials, a practice that will certainly be present in future meetings. We are sure that this meeting series will continue to provide a valuable venue for people working in simulational physics to exchange ideas and discuss the state of art of this always expanding field.

We are very glad to put together this special issue, and are most appreciative of the efforts of the editors of the *Journal of Physics: Conference Series* for making this publication possible. We are grateful for the outstanding work of the Natal team, specially from the *International Institute of Physics (IIP)*, with its highly competent professional staff who made this Meeting a pleasant one. For the first time all talks have been recorded and are free to be seen from the IIP website at https://www.iip.ufrn.br/eventsdetail.php?inf===QTUVUe. Very special thanks go also to the financial support of the Brazilian agencies CAPES and CNPq, and to the Federal Universities UFPB, UFRN, UFMG and UFMT. At last, but not least, we would like to acknowledge all of the authors for their written submissions.

Bismarck Vaz da Costa
Paulo Henrique Martins
Lucas Alvares Mol
Luiz Felipe C. Pereira
João Antonio Plascak

Guest Editors

IOP Conf. Series: Journal of Physics: Conf. Series **1012** (2018) 011002 doi:10.1088/1742-6596/1012/1/011002

Peer review statement

All papers published in this volume of *Journal of Physics: Conference Series* have been peer reviewed through processes administered by the proceedings Editors. Reviews were conducted by expert referees to the professional and scientific standards expected of a proceedings journal published by IOP Publishing.

IOP Conf. Series: Journal of Physics: Conf. Series **1012** (2018) 012001 doi:10.1088/1742-6596/1012/1/012001

20 years of the Brazilian Meeting on Simulational Physics

B. V. Costa[1] **and J. A. Plascak**[2,3]

[1]Departamento de Física, Instituto de Ciências Exatas, Universidade Federal de Minas Gerais, C.P. 702, 30123-970 Belo Horizonte, MG - Brazil
[2] Universidade Federal da Paraíba, Centro de Ciências Exatas e da Natureza - Campus I, Departamento de Física - CCEN Cidade Universitária 58051-970 - João Pessoa, PB - Brazil
[3]Center for Simulational Physics, University of Georgia, 30602 Athens, GA - USA

E-mail: bvc@fisica.ufmg.br,pla@hal.physast.uga.edu

Abstract. The history behind 20 years of the Brazilian Meeting on Simulational Physics is briefly summarized, where the first version of the event took place in 1997 and the ninth one in 2017.

The idea of holding in Brazil an international meeting on simulation in physics was born in the fall of 1996, when one of us (BVC) was on a sabbatical year at the Center for Simulational Physics (CSP) in The University of Georgia, in Athens USA. The format was chosen to be as informal as possible, following the annual Workshops at CSP - Recent Developments in Computer Simulation Studies in Condensed Matter Physics - organized by Prof. David Landau. With this in mind, in early August 1997, at the Universidade Federal de Minas Gerais, in Belo Horizonte, Minas Gerais, Brazil, the first Brazilian Meeting on Simulational Physics (BMSP) took place, organized by Profs. B.V. Costa, D.P. Landau and A.T. Bernardes. It gathered more than 50 people, including foreign and domestic speakers, graduate and undergraduate students. The annals of this meeting were published in International Journal of Modern Physics C [1]. It was indeed a success, and a second meeting was immediately scheduled for August 1999. We intended to keep the meeting as a biannual event. It should be said that these two first meetings were organized in an artisan way, with extremely low budget (as the following ones), in Belo Horizonte, Minas Gerais.

The venue of the III, IV and V BMSP, which were held in 2003, 2005 and 2007, respectively, was the beautiful colonial city of Ouro Preto, still in the state of Minas Gerais. The attendance in these meetings had been considerably increased to about 70-80 people at the peak of the events.

Since Brazil hosted the CCP2008 (the IUPAP sponsored event on computational physics), the VI BMSP was held only in 2011, and for the first time out of the mountains of Minas Gerais. The chosen venue was the Pantanal region, in the state of Mato Grosso. It was also an attempt at taking such an important meeting to more remote Brazilian regions in order to enhance and motivate the computational physics community in those areas.

The experience and success of moving the meeting to different regions encouraged researchers from other universities to take the event to their places. So, the VII, VIII and IX meetings have been organized in João Pessoa (in the state of Paraíba), Florianópolis (in the state of Santa

Figure 1: A talk by Prof. David Landau on the I BMSP back in 1997. Belo Horizonte, Minas Gerais, Brazil.

Figure 2: Participants of the III BMSP in 2003.

Catarina) and Natal (in the state of Rio Grande do Norte).

Moments of our meetings were captured in the photos shown in Figures 1 to 8. Some websites of the previous events are still on, and can be reached via the addresses listed in [2]. More information about the previous meetings can also be obtained from the foreword of the published proceedings listed in references [1, 3, 4, 5, 6, 7, 8].

1. Invited Speakers

The Table below lists the invited speakers of the Brazilian Meeting on Simulational Physics since its inauguration in 1997.

Figure 3: Participants of the IV BMSP in 2005.

Figure 4: Participants of the V BMSP in 2007.

Acknowledgments

Of course, all these meetings benefitted from the collaboration of several people who dedicated themselves deeply to bring about a highly successful event. It would be almost impossible to express our immense gratitude to all of them herein. However, we would like very much to address some special thanks to Prof. D. Landau, form CSP-UGA, who always provided us with his presence, excellent talks and suggestions for speakers. We could not forget Prof. Américo Bernardes from UFOP, without whom it would be certainly impossible to organize the initial meetings, specially in Ouro Preto. Prof. Paulo Murilo Oliveira is also gratefully acknowledged and Prof. Steve Lewis for his help with the manuscript. Concerning the financial support, we would like to thank the Brazilian agencies FAPEMIG, CAPES and CNPQ, as well as the federal

[redacted]

Figure 5: Participants of the VI BMSP in 2011. The head text is in Portuguese, as it appears in the website. Some of the speakers are indicated by the arrows.

Figure 6: Participants of the VII BMSP in 2013.

universities UFMG, UFOP, UFMT, UFPB, UFRN (so far).

References

[1] Costa B V, Landau D P and Bernardes A T 1998 *Int. J. Mod. Phys. C* **9** iii.

[2] V BMSP at http://www.bmsp.ufop.br; VII BMSP at http://www.fisica.ufpb.br/bmsp; IX BMSP at https://www.iip.ufrn.br/eventsdetail.php?inf===QTUVUe.

[3] Plascak J A and Dickman R 2004 *Braz. J. Phys.* **34** 335.

[4] Plascak J A and Dickman R 2006 *Braz. J. Phys.* **36** 1.

[5] Plascak J A and Dickman R 2008 *Braz. J. Phys.* **38** 1.

[6] Plascak J A 2012 *Int. J. Mod. Phys. C* **23** 1202001.

[7] Plascak J A and Rosas A 2014 *J. Phys.: Conf. Ser.* **487** 011001.

[8] Branco N S et al. 2016 *J. Phys.: Conf. Ser.* **686** 011001.

Figure 7: Participants of the VIII BMSP in 2015.

Figure 8: Participants of the IX BMSP in 2017.

IOP Conf. Series: Journal of Physics: Conf. Series **1012** (2018) 012001 doi:10.1088/1742-6596/1012/1/012001

Table 1: List of the invited speakers of BMSP.

MEETING	INVITED SPEAKERS
I (1997) Belo Horizonte	D. Landau (USA), D. Stauffer (Germany), P. M. C. Castro (Brazil), T. J. P. Penna (Brazil), R. M. Zorzenon dos Santos (Brazil), C. S. Shida (Brazil), V. B. Henriques (Brazil), S. Moss de Oliveira (Brazil), T. G. Rappoport (Brazil), E. Miranda (Brazil), I. L. Menezes-Sobrinho (Brazil), A. Bernardes (Brazil), P. Z. Coura (Brazil), O. N. Mesquita (Brazil), P. R. A Campos (Brazil), J. S. Sá Martins (Brazil), A. R. Lima (Brazil), A. R. R. Papa (Brazil), C. Moukarzel (Brazil)
II (1999) Belo Horizonte	Shan-Ho Tsai (USA), B. V. Costa (Brazil), J. G. Moreira (Brazil) J. Kamphorst L. Silva (Brazil), T. J. P. Penna (Brazil), P. M. C. Oliveira (Brazil), A. Bernardes (Brazil), R. Dickman (Brazil) (unfortunately, incomplete list)
III (2003) Ouro Preto	D. Landau (USA), K. Kaski (Finland), Y. Okabe (Japan), F. Aarão Reis (Brazil), Shan-Ho Tsai (USA), C. Scherer (Brazil), M. R. Braga (Brazil), S. Canuto (Brazil), W. Figueiredo (Brazil), M. Troyer (Switzerland), R. Bhatt (USA), R. Dickman (Brazil), U. H. E. Hansmann (Germany), J. P. R. (Brazil), M. de Oliveira (Brazil)
IV (2005) Ouro Preto	D. Landau (USA), P. M. C. Oliveira (Brazil), Shan-Ho Tsai (USA), V. B. Henriques (Brazil), N. Ito (Japan), R. Ziff (USA), H. Herrmann (Switzerland), K. Kremer (Germany), W. Janke (Germany), M. Novotny (USA), S. Lewis (USA), J. Kamphorst (Brazil), R. Dickman (Brazil)
V (2007) Ouro Preto	D. Landau (USA), M. Lyra (Brazil), Shan-Ho Tsai (USA), R. Almeida (Brazil), H. Herrmann (Switzerland), W. Paul (Germany), R. Dickman (Brazil), H. A. de Raedt (USA), A. Antonelli (Brazil), E. Luijten (Netherlands), D. Seaton (USA), T. Tome (Brazil)
VI (2011) Pantanal	D. Landau (USA), J. Yeomans (England), M. Novotny (USA), N. Ito (Japan), A. Thornton (USA), W. Figueiredo (Brazil), M. Barbosa (Brazil), R. Santos (Brazil), A. Bernardes (Brazil), P. M. Oliveira (Brazil), G. Krein (Brazil), A. Milchev (Bulgaria), B. V. da Costa (Brazil), R. Dickman (Brazil), M. Lyra (Brazil), D. Seaton (USA), M. Weigel (Germany), J. Rino (Brazil), S. Gonzalez (Netherlands)
VII (2013) João Pessoa	D. Landau (USA), C. Mc Cabe (USA), A. Bernardes (Brazil), M. Bachmann (USA), J. Adler (Israel), H. Herrmann (Switzerland), Y. Okamoto (Japan), P. Campos (Brazil), B. V. Costa (Brazil, D. Perera (USA), A. Lima (Brazil), A. Malakis (Greece), S. Azevedo (Brazil), N. Ito (Japan), N. Wilding (England), S.-H. Tsai (USA), S. Ferreira (USA), P. Martins (Brazil), F. Schmid (Germany)
VIII (2015) Florianópolis	D. Landau (USA), B. Liewehr (USA), T. Prellberg (England), M. Bachmann (USA), R. Zorzenon (Brazil), J. G. Ramos (Brazil), J. Rocha (Brazil), M. Barbosa (Brazil), S. Lewis (Brazil), A. Valentim (Brazil), T. Schilling (Luxemburg), S. Boettcher (USA), L.F. Pereira (Brazil), N. Ito (Japan), S.-H. Tsai (USA), R. Dickman (Brazil), M. Ostilli (Brazil), G Heinzelman (Brazil)
IX (2017) Natal	D. P. Landau (USA), F. Family (USA), L. F. C. Pereira (Brazil), U. Slems (Germany), E. D. Leonel (Brazil), R. M. Almeida (Brazil), F. W. S. Lima (Brazil), B. V. Costa (Brazil), W. Krauth (France), Y.-W. Li (USA), A. Monastra (Argentina), L. A. Mol (Brazil), M. Barbosa (Brazil), T. Schilling (Germany), P. R. A. Campos (Brazil), S. Pilati (Italy), N. Ito (Japan), M. Bachmann (USA), K. Kremer (Germany), T. Macri (Brazil), L. Machado (Brazil), A. Farris (USA)

IOP Conf. Series: Journal of Physics: Conf. Series **1012** (2018) 012002 doi:10.1088/1742-6596/1012/1/012002

92 Years of the Ising Model: A High Resolution Monte Carlo Study

Jiahao Xu[1], Alan M. Ferrenberg[2], David P. Landau[1]

1 Center for Simulational Physics, University of Georgia, Athens, GA 30602 USA

2 Information Technology Services and Department of Chemical, Paper & Biomedical Engineering, Miami University, Oxford, OH 45056 USA

E-mail: `jiahaoxu@uga.edu`, `alan.ferrenberg@miamioh.edu`, `dlandau@physast.uga.edu`

Abstract. Using extensive Monte Carlo simulations that employ the Wolff cluster flipping and data analysis with histogram reweighting and quadruple precision arithmetic, we have investigated the critical behavior of the simple cubic Ising model with lattice sizes ranging from 16^3 to 1024^3. By analyzing data with cross correlations between various thermodynamic quantities obtained from the same data pool, we obtained the critical inverse temperature $K_c = 0.221\,654\,626(5)$ and the critical exponent of the correlation length $\nu = 0.629\,912(86)$ with precision that improves upon previous Monte Carlo estimates.

1. Introduction

The Ising Model [1] has long played a special role in the theory of phase transitions and has served as a "fruit fly" model for testing new numerical and theoretical approaches. While Ising published the exact solution in 1-dim in 1925 [1] and Onsager the 2-dim version in 1944 [2], 3-dim Ising models have defied solution. Monte Carlo simulations [3], non-equilibrium relaxation Monte Carlo [4], Monte Carlo renormalization group [5, 6], field theoretic methods [7, 8] and high-temperature series expansions [9] have provided precise values for the location of the phase transition [10] and critical exponents, although the results did not all agree within the error bars. Simulations were insufficient to test Rosengren's "exact conjecture" for the critical temperature of the simple cubic Ising model [11]; moreover, Fisher showed that other "exact conjectures" gave quite similar values [12]. (For a nice review of results prior to 2002 see Ref. [10].)

Recent developments have renewed interest in the critical behavior of the 3d Ising model. The conformal bootstrap method provided precise estimates for the critical exponent ν [13, 14, 15], and Monte Carlo simulations [16, 17, 18] and tensor renormalization group theory with high-order singularity value decomposition [19] also yielded very precise results. Although Zhang claimed to have solved the 3d Ising model exactly [20], convincing arguments showed that his solution was incorrect [21, 22, 23].

We now present new results of high-precision Monte Carlo simulations of critical behavior of the simple cubic Ising model, analyzed using histogram reweighting techniques [24, 25], cross correlation analysis [26, 27] and finite-size scaling methods [28, 29, 30, 31].

2. Model and methods

2.1. Three-dimensional Ising model

We have considered the simple cubic, ferromagnetic Ising model with nearest-neighbor interactions on $L \times L \times L$ lattices with periodic boundary conditions. Each of the lattice sites i has a spin, $\sigma_i = \pm 1$ which interact with Hamiltonian

$$\mathcal{H} = -J \sum_{\langle i,j \rangle} \sigma_i \sigma_j, \tag{1}$$

where $\langle i, j \rangle$ denotes distinct pairs of nearest-neighbors and J is the interaction constant. The dimensionless energy E is

$$E = -\sum_{\langle i,j \rangle} \sigma_i \sigma_j. \tag{2}$$

The dimensionless coupling constant $K = J/k_B T$ will be useful for discussing critical behavior.

2.2. Monte Carlo algorithm

We sampled states using the Wolff cluster flipping algorithm [32] in which single clusters are grown and flipped sequentially. Clusters are created by drawing bonds to all nearest neighbors of the growing cluster with probability

$$p = 1 - e^{-2K\delta_{\sigma_i \sigma_j}} \tag{3}$$

Random numbers were produced using the Mersenne Twister generator [33]. No difference was discernible between data obtained using 32-bit and 53-bit versions. Simulations were performed at $K_0 = 0.221\,654$ (the estimate for the critical inverse temperature K_c by MCRG analysis [6], also used in an earlier, high resolution Monte Carlo study [3]). Lattice sizes were in the range $16 \leq L \leq 1024$, and measurements were taken for each run after 2×10^5 Wolff steps were discarded for equilibrium. Even for the largest lattice, $L = 1024$, the energy had reached equilibrium by $130,000$ cluster steps, and the simulation was then run at least another ten times the equilibrium relaxation time before data accumulation began. We performed 6,000 to 12,000 runs of 5×10^6 measurements for each L using a total of around 2×10^7 CPU core hours and producing more than 5TB data.

2.3. Methods of analysis

To locate the peaks in response functions precisely we used histogram reweighting [24, 25] to transform a probability distribution measured at one coupling to one at an adjacent value of coupling. After N configurations are generated at inverse temperature K_0 with a probability proportional to the Boltzmann weight, $\exp(-K_0 E)$, the probability of simultaneously observing the system with total (dimensionless) energy E and total magnetization M is estimated,

$$P_{K_0} = \frac{1}{Z(K_0)} W(E, M) \exp(K_0 E) \approx H(E, M)/N, \tag{4}$$

where $Z(K_0)$ is the partition function and $H(E, M)$ is the measured histogram of states. The (estimated) probability distribution for arbitrary K is then

$$P_K(E, M) \approx \frac{H(E, M)e^{\Delta K E}}{\sum H(E, M)e^{\Delta K E}}, \tag{5}$$

where $\Delta K = K_0 - K$ and the average value of any function of E and M, $f(E, M)$,

$$\langle f(E, M) \rangle_K = \sum_{E,M} f(E, M) P_K(E, M) \tag{6}$$

Ferrenberg and Landau [3] showed that the critical exponent ν of the correlation length can be estimated more precisely from Monte Carlo simulation data if multiple quantities, including traditional quantities which still have the same critical properties, are included. The logarithmic derivative of any power of the magnetization

$$\frac{\partial \ln \langle |m|^i \rangle}{\partial K} = \frac{1}{\langle |m|^i \rangle} \frac{\partial \langle |m|^i \rangle}{\partial K} = \frac{\langle |m|^i E \rangle}{\langle |m|^i \rangle} - \langle E \rangle, \tag{7}$$

for $i = 1, 2, ...$, can yield an estimate for ν. We also included (reduced) magnetization cumulants U_{2i} [34] defined by

$$U_{2i} = 1 - \frac{\langle |m|^{2i} \rangle}{3\langle |m|^i \rangle^2}, \quad i = 1, 2, 3, ... \tag{8}$$

whose derivatives with respect to K can also be used to estimate ν. The inverse critical temperature $K_c(L)$ can be estimated from the locations of the peaks in the above quantities as well as the specific heat

$$C = K^2 L^{-d} (\langle E^2 \rangle - \langle E \rangle^2), \tag{9}$$

the derivative of $|m|$ with respect to K,

$$\frac{\partial \langle |m| \rangle}{\partial K} = \langle |m| E \rangle - \langle |m| \rangle \langle E \rangle, \tag{10}$$

the finite-lattice susceptibility,

$$\chi' = K L^d (\langle |m|^2 \rangle - \langle |m| \rangle^2), \tag{11}$$

and the zero of the fourth-order energy cumulant

$$Q_4 = 1 - \frac{\langle (E - \langle E \rangle)^4 \rangle}{3\langle (E - \langle E \rangle)^2 \rangle^2}. \tag{12}$$

Note that Eq. (11) is for the finite-lattice susceptibility, not the "true" susceptibility calculated from the variance of m, $\chi = K L^d (\langle m^2 \rangle - \langle m \rangle^2)$ which has no peak for finite L. (For long runs, $\langle m \rangle = 0$ for $H = 0$ so that any peak in χ is merely due to finite run length.) All quantities were calculated using the GCC Quad-Precision Math Library (quadruple, 128 bit) precision.

The critical behavior of a system near a 2nd order transition can be extracted from the size dependence of the singular part of the free energy density [28, 29, 30, 31]. As a function of L (linear dimension) and T (temperature) as variables, the free energy of a system is described by the finite size scaling (FSS) ansatz,

$$F(L, T) = L^{-(2-\alpha)/\nu} \mathcal{F}(\varepsilon L^{1/\nu}, h L^{(\gamma+\beta)/\nu}), \tag{13}$$

where $\varepsilon = (T - T_c)/T_c$ (T_c is the infinite-lattice critical temperature), h is the magnetic field, and the critical exponents α, β, γ and ν are the infinite lattice values. Various thermodynamic properties can be determined from Eq. (13) and have corresponding scaling forms, e.g.,

$$m = L^{-\beta/\nu} \mathcal{M}^0 (\varepsilon L^{1/\nu}), \tag{14}$$

$$\chi = L^{\gamma/\nu} \chi^0 (\varepsilon L^{1/\nu}), \tag{15}$$

$$C = L^{\alpha/\nu} \mathcal{C}^0 (\varepsilon L^{1/\nu}), \tag{16}$$

where $\mathcal{M}^0(x)$, $\chi^0(x)$ and $\mathcal{C}^0(x)$ are scaling functions. Because we are interested in zero-field properties ($h = 0$), x is the only relevant thermodynamic variable. Then, for finite L, e.g. the specific heat peak occurs where the scaling function $\mathcal{C}^0$ is maximum, i.e., when

$$\left.\frac{\partial \mathcal{C}^0(x)}{\partial x}\right|_{x=x^*} = 0. \tag{17}$$

The temperature corresponding to the peak is the finite-lattice "transition" temperature $T_c(L)$, on the condition $x = x^*$ which varies with L asymptotically as

$$T_c(L) = T_c + T_c x^* L^{-1/\nu}. \tag{18}$$

The FSS ansatz is valid only for sufficiently large L and T sufficiently close to T_c. Corrections to scaling and finite-size scaling terms appear for smaller systems and temperatures away from T_c due to: irrelevant scaling fields which can be expressed in terms of an exponent θ, e.g. $a_1\varepsilon^\theta + a_2\varepsilon^{2\theta} + \cdots$, and others due to the non-linear scaling fields, e.g. $b_1\varepsilon^1 + b_2\varepsilon^2 + \cdots$. The correction terms can then be expressed by: $a_1L^{-\theta/\nu} + a_2L^{-2\theta/\nu}$ and $b_1L^{-1/\nu} + b_2L^{-2/\nu}$. Including correction terms, the estimate for the inverse critical coupling $K_c(L)$ becomes

$$K_c(L) = K_c + A_0L^{-1/\nu}(1 + A_1L^{-\omega_1} + A_2L^{-\omega_2} + \cdots). \tag{19}$$

We first estimate the critical exponent ν and then insert it into Eq. (19), so that there is one less unknown parameter to do the non-linear fit to Eq. (19). To do this we can use the critical scaling form without the prior knowledge of the transition coupling K_c

$$\left.\frac{\partial U_{2i}}{\partial K}\right|_{\max} = U_{i,0}L^{1/\nu}(1 + a_1L^{-\omega_1} + a_2L^{-\omega_2} + \cdots) \tag{20}$$

$$\left.\frac{\partial \ln \langle |m|^i \rangle}{\partial K}\right|_{\max} = D_{i,0}L^{1/\nu}(1 + a_1L^{-\omega_1} + a_2L^{-\omega_2} + \cdots) \tag{21}$$

Once ν is determined from the fit of Eq. (20) and Eq. (21), we can estimate the critical inverse temperature K_c with a fixed value of ν. K_c can also be estimated from Binder's 4th order cumulant crossing technique [34]. As $L \to \infty$, the fourth-order magnetization cumulant $U_4 \to 0$ for $K < K_c$ and $U_4 \to 2/3$ for $K > K_c$. U_4 can be plotted as a function of K for different lattice sizes, and the location of the intersections between curves for the two lattice sizes is given by

$$K_{\text{cross}}(L, b) = K_c + a_1L^{-1/\nu-\omega_1}\left(\frac{b^{-\omega_1} - 1}{b^{1/\nu} - 1}\right) + a_2L^{-1/\nu-\omega_2}\left(\frac{b^{-\omega_2} - 1}{b^{1/\nu} - 1}\right) + \cdots$$

where L is the size of the smaller lattice, $b = L'/L$ is the ratio of two lattice sizes, and ω_1, ω_2 are correction exponents in the finite-size scaling formulation.

Ideally, each configuration only depends on the previous configuration, but, in practice, it is likely to be correlated to earlier configurations. The correlation decays with separation in Monte Carlo time, but fluctuations appear smaller than they should be. To deal with this issue, we can consider blocks of the original data, and use jackknife resampling [35].

An important advance by Weigel and Janke [26, 27] was the seminal observation that cross correlation between different quantities could lead to systematic bias in the estimates of critical quantities extracted from the data. For a set of n measurements of a random variable $\mathbf{x} = (x_1, x_2, \cdots, x_n)$, and an estimator $\hat{\theta} = f(\mathbf{x})$, the Jackknife estimate of the value and error of $\hat{\theta}$ is found by leaving out one measurement at a time. We define the jackknife average, x_i^J by,

$$x_i^J = \frac{1}{n-1}\sum_{j \neq i} x_j, \tag{22}$$

where $i = 1, 2, \cdots, n$, so x_i^J is the average of all the x values except x_i. Similarly, we define

$$\hat{\theta}_i^J = f(x_i^J). \tag{23}$$

The jackknife estimate of $\hat{\theta} = f(\mathbf{x})$ is the average of $\hat{\theta}_i^J$, i.e.

$$\bar{\theta} = \frac{1}{n}\sum_{i=1}^{n} \hat{\theta}_i^J = \frac{1}{n}\sum_{i=1}^{n} f(x_i^J), \tag{24}$$

and the jackknife error $\sigma(\hat{\theta})$, is given by,

$$\sigma(\hat{\theta}) = \left[\frac{n-1}{n}\sum_{i=1}^{n}(\hat{\theta}_i^J - \bar{\theta})^2\right]^{1/2}. \tag{25}$$

In general there can be multiple adjacent elements in each block, and significant cross correlations may exist between estimates $\hat{\theta}^{(k)}$ and $\hat{\theta}^{(l)}$ from the same original time-series data, To reduce these effects, we calculated the jackknife covariance matrix $\mathbf{G} \in \mathbb{R}^{m \times m}$ [35]. For a number of estimates $\hat{\theta}^{(k)}$, the rth row, c^{th} column entry of matrix $\mathbf{G}$ is given by,

$$\mathbf{G}_{rc}(\hat{\theta}) = \frac{n-1}{n}\sum_{i=1}^{n}(\hat{\theta}_i^{J,(r)} - \bar{\theta}^{(r)})(\hat{\theta}_i^{J,(c)} - \bar{\theta}^{(c)}). \tag{26}$$

The m different estimates $\hat{\theta}^{(k)} (k = 1, 2, ..., m)$ for the same parameter $\hat{\theta}$, should have the same expectation value. So the estimated value for $\hat{\theta}$ can be determined by a linear combination,

$$\bar{\theta} = \sum_{k=1}^{m} \alpha_k \hat{\theta}^{(k)}. \tag{27}$$

where $\sum_k \alpha_k = 1$. Based on the cross correlation analysis from Ref. [26, 27], with the optimal choice for the weights, the variance can be expressed by,

$$\sigma^2(\hat{\theta}) = \frac{1}{\sum_{k=1}^{m}\sum_{l=1}^{m}[\mathbf{G}(\hat{\theta})^{-1}]_{kl}}. \tag{28}$$

At our level of resolution the finite precision of the pseudorandom number generator and the numbers that can be stored in memory become relevant. When the Wolff algorithm probability of adding a spin to the cluster is converted to a 32-bit unsigned number for comparison with a pseudorandom number generated in the simulation, it is truncated due to finite precision. When reconverted back into a value of K the result differs from 0.221 654 in the 10th decimal place. For the largest L, this is only a factor of 20 smaller than the statistical error. Simulations with a 53-bit pseudorandom number generator showed that this is not significant for our analysis, but for future studies of larger systems and/or higher precision, a 32-bit random number generator might not suffice. We also used the corrected, effective K_0 instead of 0.221654 and for $L = 1024$ we used the multiple-histogram method [25] to combine results for the 32 and 53-bit pseudorandom number generators.

Comparisons of results from Wolff cluster flipping runs and Metropolis single spin-flip runs for $L = 32$ showed no differences. The Wolff cluster simulation for L=32 was repeated using the MRG32K3A random number generator from Pierre L'Ecuyer, "Combined Multiple Recursive Random Number Generators", Operations Research, 47, 1 (1999), 159-164. The locations and values of the maxima in all quantities agreed, to within the error bars, as those from the Metropolis simulations and the Wolff simulations with the Mersenne Twister. Hence the problems found by Ferrenberg et al. [36] using other random number generators were not noticeable here.

Table 1. Estimates for ν with: one fixed correction exponent $\omega_1 = 0.83$, two fixed exponents $\omega_1 = 0.83$, $\omega_2 = 4$, and three fixed exponents $\omega_1 = 0.83$, $\omega_2 = 4$, $\omega_3 = 1.6$ as a function of $L_{\min}$.

$L_{\min}$	$\nu(\omega_1$ fixed)	$\nu(\omega_{1,2}$ fixed)	$\nu(\omega_{1,2,3}$ fixed)
16	0.631 814(18)	0.630 806(30)	0.630 072(45)
24	0.631 046(26)	0.630 513(40)	0.630 049(57)
32	0.630 722(33)	0.630 241(55)	0.629 980(77)
48	0.630 350(48)	0.630 278(78)	0.629 99(11)
64	0.630 319(62)	0.630 21(11)	0.630 06(15)
80	0.630 285(78)	0.630 10(15)	0.629 93(21)
96	0.630 25(10)	0.629 93(18)	0.629 90(29)
112	0.630 14(13)	0.630 01(17)	0.629 93(18)
128	0.630 04(15)	0.630 04(15)	0.629 84(22)
144	0.629 85(18)	0.629 85(18)	0.629 96(26)

3. Results

3.1. Finite-size scaling analysis

The conformal bootstrap [13, 14] gives a theoretical prediction for the confluent correction exponents,

$$\omega_1 = 0.8303(18), \quad \omega_2 \approx 4. \tag{29}$$

and the correction term corresponding to the non-linear scaling fields is [37],

$$\omega_3 = 1/\nu \tag{30}$$

Thus, the fitting model for the peak value of quantity X is,

$$X_{\max} = X_0 L^{1/\nu}(1 + a_1 L^{-\omega_1} + a_2 L^{-\omega_2} + a_3 L^{-\omega_3}) \tag{31}$$

We analyzed the data fitting model Eq. (31) using between one and three correction terms with fixed powers $\omega_1 = 0.83$, $\omega_2 = 4$, $\omega_3 = 1.6$. By calculating the covariance matrix and doing the cross-correlation analysis, we give estimates for ν in Table 1 where the minimum lattice size included in the analysis, $L_{\min}$, is eliminated one by one. With three correction exponents, two for confluent corrections ($\omega_1 = 0.83$, $\omega_2 = 4$) and one for the non-linear scaling fields ($\omega_3 = 1.6$), the estimate for the critical exponent ν shows statistical fluctuations; but the value for ν was estimated by taking the average of ν obtained from fits for different $L_{\min}$ and the error estimated with the jackknife method on estimates of ν from the analysis using different ranges of relatively large $L_{\min}$. Using values of $L_{\min}$ from 80 to 144, we find

$$\nu = 0.629\,912(86). \tag{32}$$

Since higher order terms like $a_{1,2}L^{-2\omega_1} + \cdots$ could be present and larger than the higher order confluent corrections, we repeated the analysis using these higher order corrections instead. The results for the asymptotic values of the critical temperature and exponent were essentially identical. Thus, either the higher order terms have small prefactors or we do not have sufficient resolution to differentiate their contribution from those of higher order confluent corrections.

To estimate the critical coupling K_c, we fitted the locations of the peak of the multiple quantities using between one and three fixed correction exponents, $\omega_1 = 0.83$, $\omega_2 = 4$, $\omega_3 = 1.6$, to the fitting model Eq. (33),

$$K_c(L) = K_c + A_0 L^{-1/\nu}(1 + A_1 L^{-\omega_1} + A_2 L^{-\omega_2} + A_3 L^{-\omega_3}). \tag{33}$$

Table 2. Fitted values for K_c with: (left column) one correction term (fixed exponent $\omega_1 = 0.83$); (center column) two correction terms (fixed exponents $\omega_1 = 0.83$, $\omega_2 = 4$; and (right column) three correction terms (fixed exponents $\omega_1 = 0.83$, $\omega_2 = 4$, $\omega_3 = 1.6$) as a function of $L_{\min}$.

$L_{\min}$	K_c(1 fixed ω)	K_c(2 fixed ω)	K_c(3 fixed ω)
16	0.221 654 656 2(10)	0.221 654 639 3(11)	0.221 654 625 7(21)
24	0.221 654 638 8(11)	0.221 654 630 8(12)	0.221 654 625 7(24)
32	0.221 654 634 3(11)	0.221 654 630 7(12)	0.221 654 625 3(32)
48	0.221 654 630 7(12)	0.221 654 630 5(12)	0.221 654 623 2(30)
64	0.221 654 628 4(13)	0.221 654 628 4(13)	0.221 654 623 4(60)
80	0.221 654 627 5(14)	0.221 654 627 5(15)	0.221 654 625 0(75)
96	0.221 654 626 0(17)	0.221 654 626 0(16)	0.221 654 627 9(97)
112	0.221 654 625 9(18)	0.221 654 626 0(18)	0.221 654 625 0(49)
128	0.221 654 625 8(21)	0.221 654 625 8(21)	0.221 654 626 3(48)
144	0.221 654 627 0(25)	0.221 654 627 0(25)	0.221 654 627 1(34)

The results for K_c are shown in Table 2.

The fluctuation of K_c when $L_{\min} \leq 80$ is larger than when $L_{\min} \geq 80$ and finite-size effects diminish for larger L. From the average of K_c for $L_{\min} = 80$ to 144 we estimate $K_c = 0.221\,654\,626\,2$. Likewise, a jackknife analysis has been done on the estimates for K_c which are obtained from the three correction terms analysis. From values of $L_{\min}$ from 80 to 144 we estimate $K_c = 0.221\,654\,626\,2(23)$ whereas for $L_{\min} = 16$ to 144, $K_c = 0.221\,654\,625\,5(42)$. Therefore, we conclude from the finite size scaling analysis, with conservative error bars, that

$$K_c = 0.221\,654\,626(5). \tag{34}$$

As seen in Fig. 2, effects of higher order correction terms are clearly visible.

3.2. Crossing technique of the 4th order magnetization cumulant

As the lattice size $L \to \infty$, the fourth-order magnetization cumulant $U_4 \to 0$ for $K < K_c$ and $U_4 \to 2/3$ for $K > K_c$. For large enough lattice sizes, curves for U_4 cross as a function of inverse temperature at a "fixed point" U^*, and the location of the crossing "fixed point" is K_c. Because of finite-size correction terms, not all curves cross at a common intersection. The locations of the cumulant crossings have been fitted to Eq. (22) with two correction terms. The critical coupling appears to be stable if $L_{\min} \geq 96$ and the average of K_c values for $L_{\min} \geq 96$ yields

$$K_c = 0.221\,654\,628(2) \tag{35}$$

3.3. Discussion

The combination of an efficient, cluster-flipping Monte Carlo algorithm, high statistics simulations, histogram reweighting, and a cross-correlation jackknife analysis enables the high resolution of the results presented above. In Fig. 3, we show the results of our Monte Carlo study as well as other high-resolution simulations estimates for ν and K_c. The boxes represent the quoted error bars in both ν and K_c assuming independent errors.

Our estimate $\nu = 0.629\,912(86)$ is consistent (i.e. within the error bars) with the recent conformal bootstrap result of Kos et al. [15], as well as that from an older work by El-Showk et al. [14]. Also, our result agrees well with the Monte Carlo result of Hasenbusch [16] but is lower

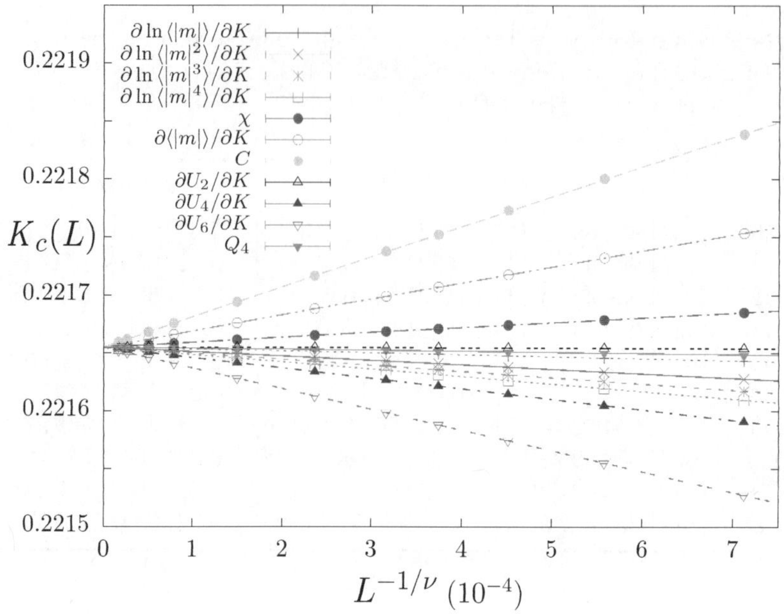

Figure 1. Size dependence of the finite-lattice effective critical temperatures estimated from $L = 96$ to 1024.

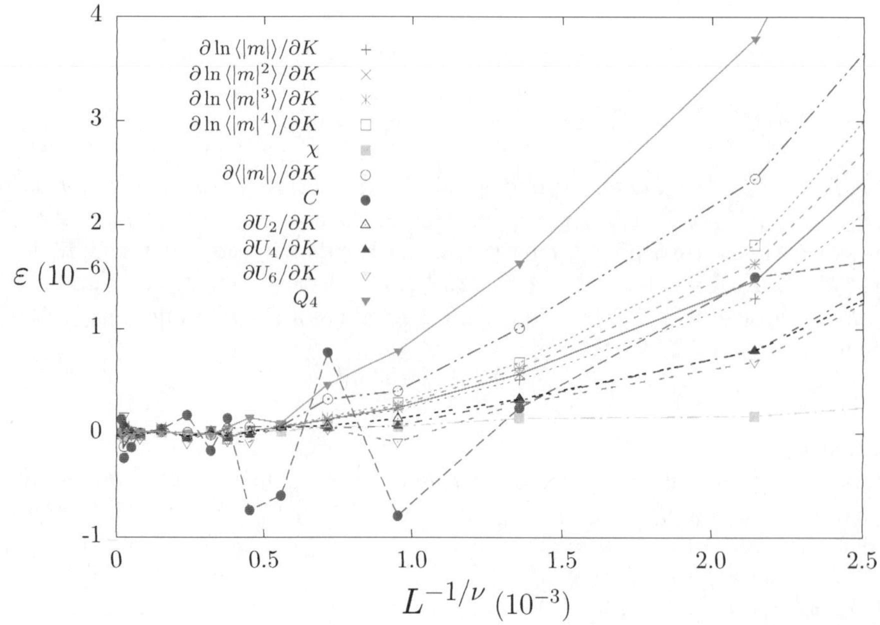

Figure 2. Size dependence of the residual difference between measured K_c and fitted values including one correction term.

Table 3. Comparison of our results for the critical coupling K_c and the critical exponents ν, γ with other recently obtained values. Numbers marked with * are calculated by Fisher's scaling law $\gamma = \nu(2-\eta)$, and the error estimates assume that ν an η are independent and uncorrelated.

Method	K_c	ν	γ
conformal bootstrap [15]		0.629 971(4)	1.237 075(8)*
MC [27]	0.221 657 03(85)	0.630 0(17)	1.240 9(62)*
MC [16]	0.221 654 63(8)	0.630 02(10)	1.237 19(21)*
MC [17]	0.221 654 604(18)		
tensor RG [19]	0.221 654 555 5(5)		
Our results (no fit assumptions), MC	0.221 654 630(7)	0.629 60(15)	1.236 41(45)
Our results (constrained fits), MC	0.221 654 626(5)	0.629 912(86)	1.236 94(31)
Our results (cumulant crossings), MC	0.221 654 628(2)		

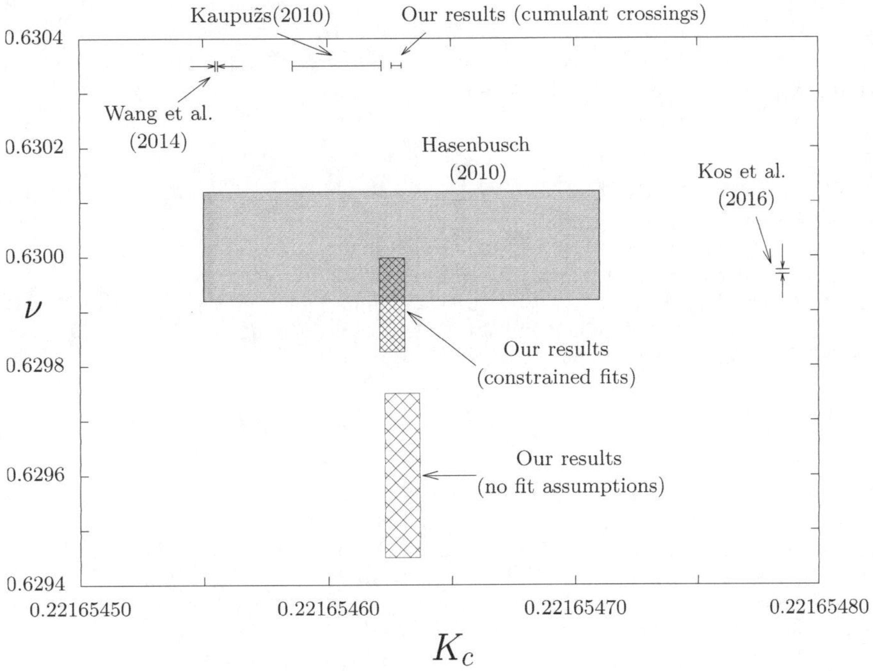

Figure 3. High-resolution estimates for K_c and ν obtained using different methods

than that of Weigel and Janke [27]; however, within the respective error bars there is agreement although we have substantially higher precision than either of these previous studies. Our system sizes and statistics are substantially greater than those used by Weigel and Janke and Hasenbusch examined the behavior of the ratio of partition functions Z_a/Z_p, and the second moment correlation length over the linear lattice size ξ_2/L so the methodologies differ. Our estimate for K_c differs from that obtained by Kaupužs et al [17] using a parallel Wolff algorithm by an amount that is slightly outside the error bars. Somewhat perplexingly, they were able to fit their data to two rather different values of ν, so no comparison of critical exponents is possible. The recent tensor renormalization group result for K_c [19] disagrees with our result by many times the respective error bars.

4. Conclusion

We have studied a 3d Ising model with the Wolff cluster flipping algorithm, histogram reweighting, and finite size scaling including cross-correlations using quadruple precision arithmetic for the analysis. Using a wide range of system sizes, with the largest containing more than 10^9 spins, and including corrections to scaling, we have obtained results for K_c, ν, and γ that are comparable in precision to those from the latest theoretical predictions and can provide independent verification of the predictions from those methods. Our values provide further numerical evidence that none of the purported "exact" values are correct. To within error bars we obtain the same value for the critical exponent ν as that predicted by the conformal bootstrap; however, our estimate for the critical temperature K_c does not agree with that from the tensor renormalization group.

As computer power increases, new sources of error must be taken into account. To improve precision more stringent tests of the random number generator and much greater statistics for much larger lattice sizes will be needed.

Acknowledgments

We thank Dr. M. Weigel and Dr. S.-H. Tsai for valuable discussions. Computing resources were provided by the Georgia Advanced Computing Resource Center, the Ohio Supercomputing Center, and the Miami University Computer Center.

References

[1] Ising E 1925 *Z. Phys.* **31** 253–258
[2] Onsager L 1944 *Phys. Rev.* **65** 117–149
[3] Ferrenberg A M and Landau D P 1991 *Phys. Rev. B* **44** 5081
[4] Ozeki Y and Ito N 2007 *J. Phys. A: Math. Theor.* **40** R149–R203
[5] Blöte H W J, Heringa J R, Hoogland A, Meyer E W and Smit T S 1996 *Phys. Rev. Lett.* **76** 2613–2616
[6] Pawley G S, Swendsen R H, Wallace D J and Wilson K G 1984 *Phys. Rev. B* **29** 4030–4040
[7] Guida R and Zinn-Justin J 1998 *J. Phys. A* **31** 8103
[8] Pogorelov A A and Suslov I M 2008 *J. Exp. Theor. Phys.* **106** 1118
[9] Butera P and Comi M 2005 *Phys. Rev. B* **72** 014442
[10] Pelissetto A and Vicari E 2002 *Phys. Rep.* **368** 549
[11] Rosengren A 1986 *J. Phys. A: Math. Gen.* **19** 1709
[12] Fisher M E 1995 *J. Phys. A: Math. Gen.* **28** 6323
[13] El-Showk S, Paulos M F, Poland D, Rychkov S, Simmons-Duffin D and Vichi A 2012 *Phys. Rev. D* **86** 025022
[14] El-Showk S, Paulos M F, Poland D, Rychkov S, Simmons-Duffin D and Vichi A 2014 *J. Stat. Phys.* **157** 869
[15] Kos F, Poland D, Simmons-Duffin D and Vichi A 2016 *J. High Energ. Phys.* **2016** 36
[16] Hasenbusch M 2010 *Phys. Rev. B* **82** 174433
[17] Kaupužs J, Rimsans J and Melnik R V N 2011 *Ukr. J. Phys.* **56** 845
[18] Kaupužs J, Melnik R V N and Rimsans J 2014 *ArXiv e-prints* (*Preprint* 1407.3095)
[19] Wang S, Xie Z Y, Chen J, Normand B and Xiang T 2014 *Chin. Phys. Lett.* **31** 070503
[20] Zhang Z 2007 *Phil, Mag.* **87** 5309
[21] Wu F, McCoy B M, Fisher M E and Chayes L 2008 *Phil. Mag.* **88** 3093–3095
[22] Wu F, McCoy B M, Fisher M E and Chayes L 2008 *Phil. Mag.* **88** 3103–3103
[23] Fisher M E and Perk J H H 2016 *Phys. Lett. A* **380** 1339 – 1340
[24] Ferrenberg A M and Swendsen R H 1988 *Phys. Rev. Lett.* **61** 2635
[25] Ferrenberg A M and Swendsen R H 1989 *Phys. Rev. Lett.* **63** 1195
[26] Weigel M and Janke W 2009 *Phys. Rev. Lett.* **102** 100601
[27] Weigel M and Janke W 2010 *Phys. Rev. E* **81** 066701
[28] Fisher M E 1971 *Critical Phenomena* (New York: Academic Press) pp 1–98
[29] Fisher M E and Barber M N 1972 *Phys. Rev. Lett* **28** 1516
[30] Barber M N 1983 *Phase Transitions and Critical Phenomena* vol 8 (New York: Academic Press) pp 146–266
[31] Privman V 1990 *Finite-Size Scaling and Numerical Simulation* (Singapore: World Scientific)
[32] Wolff U 1989 *Phys. Rev. Lett.* **62** 361–364
[33] Matsumoto M and Nishimura T 1998 *ACM Trans. Model. Comput. Simul.* **8** 3–30 ISSN 1049-3301
[34] Binder K 1981 *Z. Phys. B* **43** 119

[35] Efron B and Tibshirani R J 1993 *An Introduction to the Bootstrap* (New York: Chapman and Hall)
[36] Ferrenberg A, Landau D P and Wong Y 1992 *Phys. Rev. Lett.* **69** 3382–3384
[37] Aharony A and Fisher M E 1983 *Phys. Rev. B* **27** 4394–4400

IOP Conf. Series: Journal of Physics: Conf. Series **1012** (2018) 012003 doi:10.1088/1742-6596/1012/1/012003

A practical guide to replica-exchange Wang–Landau simulations

Thomas Vogel[1], Ying Wai Li[2] and David P Landau[3]

[1] Department of Physics, University of North Georgia, Dahlonega, GA 30597, USA

[2] National Center for Computational Sciences, Oak Ridge National Laboratory, Oak Ridge, TN 37831, USA *

[3] Center for Simulational Physics, The University of Georgia, Athens, GA 30602, USA

E-mail: `tvogel@ung.edu, yingwaili@ornl.gov, dlandau@physast.uga.edu`

Abstract. This paper is based on a series of tutorial lectures about the replica-exchange Wang–Landau (REWL) method given at the IX Brazilian Meeting on Simulational Physics (BMSP 2017). It provides a practical guide for the implementation of the method. A complete example code for a model system is available online. In this paper, we discuss the main parallel features of this code after a brief introduction to the REWL algorithm. The tutorial section is mainly directed at users who have written a single-walker Wang–Landau program already but might have just taken their first steps in parallel programming using the Message Passing Interface (MPI). In the last section, we answer "frequently asked questions" from users about the implementation of REWL for different scientific problems.

1. Introduction

Since its introduction in 2013 [1, 2], the replica-exchange Wang–Landau (REWL) method has been used to study a variety of problems, amongst them the self-assembly of lipid bilayer membranes [3, 4], polymer adsorption in the hydrophobic-polar (HP) model [5], magnetic phase transition in coupled spin-lattice systems [6], higher-dimensional parameter space sampling [7], fully-connected magnetic models of dipoles [8], network reliability [9, 10], stochastic analytic continuation of quantum Monte Carlo data [11], or ferroelectric phase transitions [12].

REWL combines the Wang–Landau (WL) method [13, 14] with elements from canonical replica-exchange methods, often termed "parallel tempering" (PT) [15, 16]. The idea was to develop a framework that is able to run on large-scale supercomputers while still leveraging the advantages of the WL method on the individual computing cores. However, we understand that there might be considerable technical barriers for the users to overcome before they can actually start to experiment with REWL. One major advantage of the original, single-walker WL method that made it so popular and robust is that it is straightforward and easy to implement. Since

*This manuscript has been co-authored by UT-Battelle, LLC under Contract No. DE-AC05-00OR22725 with the U.S. Department of Energy. The United States Government retains and the publisher, by accepting the article for publication, acknowledges that the United States Government retains a non-exclusive, paid-up, irrevocable, world-wide license to publish or reproduce the published form of this manuscript, or allow others to do so, for United States Government purposes. The Department of Energy will provide public access to these results of federally sponsored research in accordance with the DOE Public Access Plan (http://energy.gov/downloads/doe-public-access-plan).

this is less so for REWL, we decided to write this paper to provide practical guidance and help. In Section 2, we first review the main ideas of REWL without going into too much detail conceptually. In Section 3 we introduce the basic functionality of the Message Passing Interface (MPI) before discussing, in great detail, essential code segments to convert a single-walker WL code into a parallel REWL version. A full REWL example code can be found online [17]. Finally, in Section 4 we answer questions concerning implementation details that were frequently asked via e-mail and during the IX Brazilian Meeting on Simulational Physics (BMSP 2017) and other meetings.

2. REWL: A brief description of the algorithm

Here we briefly outline the Replica–Exchange Wang–Landau (REWL) algorithm. For a detailed description of the algorithm, readers are encouraged to refer to the original publications [1, 2] and the corresponding conference proceedings [3, 5]. To begin with, the first step is to determine the relevant energy range over which the physical system can range, as in Wang–Landau sampling. This total energy range is then divided into multiple, smaller, overlapping windows (Figure 1). This approach is similar to an early attempt to parallize Wang–Landau sampling [18], with two

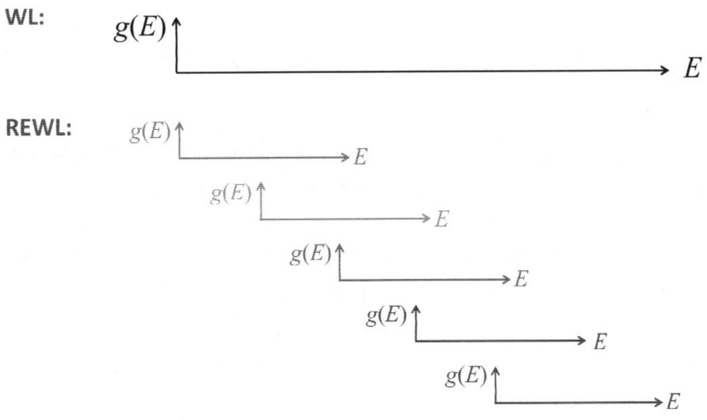

Figure 1. A schematic diagram for replica–exchange Wang–Landau sampling, Each of the multiple, overlapping energy windows covers a part of the entire energy range in the original Wang–Landau sampling.

major differences: in the previous scheme, the simulations for different energy windows were independent and could be run at different times. In REWL, a single simulation takes place with all energy windows running individual Wang–Landau sampling processes at the same time. This is necessary because at fixed intervals, i.e., every fixed number of Monte Carlo steps, replica-exchanges between the windows are performed. The probability of accepting a replica exchange, $P(X \leftrightarrow Y)$, of configurations X and Y between walkers i and j is given by:

$$P(X \leftrightarrow Y) = \min \left[1, \frac{g_i(E_X)}{g_i(E_Y)} \frac{g_j(E_Y)}{g_j(E_X)} \right], \tag{1}$$

where $g_i(E_X)$ is the current estimator for the density of states of walker i at the energy of its configuration X.

Every energy window performs one or more ordinary Wang–Landau sampling processes with their own sets of histogram and density of states, except for the occasional replica-exchanges. Since each of them proceeds with its own “pace” and could be at a different stage towards convergence, it should be noted that the simulation terminates only when the “slowest” window has attained convergence, i.e, the pre-defined value of the modification factor is reached. Before that, every energy windows has to keep performing Wang-Landau sampling even if its own density of states has converged, so as to facilitate the replica-exchanges with the windows that have not finished.

Upon completion, REWL yields a number of fragments of overlapping density of states. As the density of states is determined up to a constant, only the relative values, or the curvatures, are meaningful at this point. To combine two fragments of density of states, say the ith and the jth fragments $g_i(E)$ and $g_j(E)$, a joining point needs to be chosen as a reference point to rescale or shift one of the density of states fragments. The inverse microcanonical temperature $\beta = d\log[g(E)]/dE$ is first calculated for the overlapping regions, i.e., the derivatives of the logarithm of the density of states. Different numerical methods taking multiple bins into account can be used for this calculation. The joining point is chosen as the energy where the β-values from the two overlapping pieces best agree, denoted by E_{join}. $g_j(E)$ is then rescaled by the difference between itself and $g_i(E)$ at E_{join}, i.e., $g_{j,\text{rescaled}}(E) = g_j(E) \cdot g_i(E_{\text{join}})/g_j(E_{\text{join}})$. Once the density of states for the entire energy range is recovered, it can be normalized in the same way as if it had been obtained by the ordinary Wang–Landau scheme. Note that this scheme has been generalized and applied to reconstruct two-dimensional or higher-dimensional densities of states [5, 7].

3. REWL: A tutorial

3.1. MPI: the Message Passing Interface

Here, we further assume that each energy window performs its own Wang–Landau sampling on a single computing unit. The replica-exchange operations then require communications between a pair of computing units. Also, each computing unit needs to communicate with all other computing units from time to time to inform them of its progress towards convergence. In practice, such communications among computing units can be implemented using a parallel programming interface, the Message Passing Interface (MPI).

MPI is a standard, or specification, for a message passing library, which defines the application programming interface (API) of a message-passing system for communications between or among computing units. These can be CPU cores or compute nodes on a workstation, for instance. MPI was originally designed for distributed memory computing architectures (such as workstations), but it can also be used on shared memory architectures (e.g. multicore CPUs on a laptop). In both cases, MPI employs a distributed memory programming model (Figure 2), which assumes that each computing unit possesses independent memory and access to it is not shared with other computing units. There are different MPI implementations, such as MPICH [19], MVAPICH [20], and OpenMPI [21], to name a few. All of them provide C/C++ and Fortran interfaces. For interested readers there are a number of useful references [22–24] that provide more detail about MPI programming.

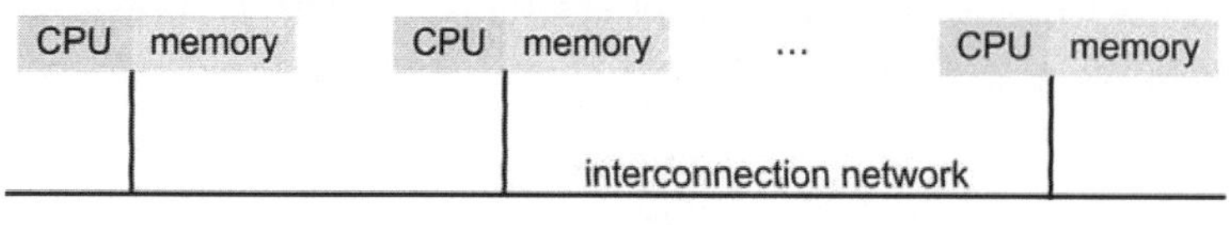

Figure 2. Message Passing Interface (MPI) assumes distributed memory programming model.

3.2. Hello World: Introduction to the necessary MPI functions

In the following sample code we demonstrate the very essential MPI functions necessary to initialize the MPI environment and to send and receive data between different computing cores in a C/C++ program.

IOP Conf. Series: Journal of Physics: Conf. Series **1012** (2018) 012003 doi:10.1088/1742-6596/1012/1/012003

```
/*** Example: basic point-to-point communication  ***/

#include <stdio.h>
#include "mpi.h"   // Include MPI header file containing the librarys API

int main (int argc, char* argv[])
{
  int my_rank, num_procs;

  // Start up MPI
  MPI_Init (&argc, &argv);

  // Get current process' "ID" number (rank)
  MPI_Comm_rank (MPI_COMM_WORLD, &my_rank);

  // Get total number of processes
  MPI_Comm_size (MPI_COMM_WORLD, &num_procs);

  // Store the status of whether a communication is successful or not
  MPI_Status status;

  printf("Hello World! This is process %d of %d\n", my_rank, num_procs);

  int message = 42;          // Content of the message
  int msg_tag = 23;          // A unique tag for each message

  if (my_rank == 0)          // Process 0 receives a message from Process 1
  {
    MPI_Recv(&message, 1, MPI_INT, 1, msg_tag, MPI_COMM_WORLD, &status);
    printf("rank 0: recved message %d from %d\n", message, status.MPI_SOURCE);
  }
  else if (my_rank == 1)    // Process 1 sends a message to Process 0
  {
    MPI_Send(&message, 1, MPI_INT, 0, msg_tag, MPI_COMM_WORLD);
    printf("I am Rank 1. Sent a message to Rank 0.\n");
  }
  else                       // All other processes sit and wait
    printf("I am Process %d. I have nothing to do.\n", my_rank);

  MPI_Finalize();            // Finish up MPI

  return 0;
}
```

- `MPI_Init` initializes MPI with the command line arguments supplied to the program's executable
- `MPI_Comm_rank` gets the rank number of the current MPI process inside the specific MPI communicator
- `MPI_Comm_size` gets the total number of processes in the specific MPI communicator
- `MPI_Finalize` finishes up and exits MPI.

More specifically, we demonstrate in this example a simple point-to-point communication where MPI rank 0 receives a message (data) containing an integer number (42) from MPI rank 1. If there are more than two MPI ranks initiated, these MPI ranks are not involved in the communication. MPI rank 0 calls `MPI_Recv`, while MPI rank 1 calls `MPI_Send`. In this example we use the global *communicator* `MPI_COMM_WORLD`, a predefined communication channel that includes all processes. Note that we will use dedicated communicators (channels) and more specialized communication functions later.

We suggest running this example code and familiarizing yourself with it. It includes all necessary concepts to understand the fundamental functionality of any MPI program. Note that the header file `mpi.h` which contains all the MPI application program interfaces (APIs) has to be included at the beginning of the program.

3.3. Parallel functions/extensions for REWL in detail

At this point we assume that the reader has a working Wang–Landau code and we discuss necessary extensions to convert that single-processor code into a parallel REWL code. While there are different ways of implementing each functionality, the basic design assumptions are the following:

- Every walker, performing ordinary Wang–Landau sampling, gets its own MPI process or rank.
- $n \times m$ MPI processes are launched (m - number of walkers in each energy window; n - number of energy windows), each with its individual energy range, histogram, and density-of-states estimator. There is no "master" process.
- All data and replica-exchanges are facilitated exclusively via MPI communications.

All code snippets shown in the following are taken from the code used to produce the data presented in [1, 2]. Please feel free to download the full code online [17].

3.3.1. Preparations: setting up MPI and communicators We will demonstrate in this section how individual communicators are set up and used instead of relying on the global communicator `MPI_COMM_WORLD` for communications. This could be advantageous for various reasons. First and foremost, this keeps the design clean and communications compartmentalized. Another benefit is that resilience measures can be added in a way that the failure of a single process or data transfer can be repaired and will not shut down the communications between uninvolved processes and eventually the whole simulation [25, 26].

Every process will be part of two separate communicators, one to communicate with all other walkers in the neighboring energy window on the left, the other to exchange data with any walker in the energy window on the right. For example, in Figure 3 walkers 4–6 would be in two communicators: one includes walkers 1–6, the other includes walkers 4–9. The communicators can be set up as follows.

```
// Set up local MPI communicators for replica-exchange (RE)
// - each process belongs to two local groups/communicators
// - each process has different local IDs in different communicators generally

int numprocs;          // total number of walkers
int myid;              // my rank(ID) in the global communicator, MPI_COMM_WORLD
int multiple;          // number of walkers having the same energy window
int comm_id;           // ID for a communicator
int mylocalid[2];      // my ID in local communicators
```

```
// Get the group of processes in MPI_COMM_WORLD (i.e., all processors)
MPI_Group world;
MPI_Comm_group(MPI_COMM_WORLD, &world);

// The followings are for defining a list of MPI local communicators

int *ranks;                                       // an array to store global IDs
ranks = (int*) malloc(2*multiple*sizeof(int)); // for walkers in a local comm.

int numLocalComm = (numprocs/multiple) - 1; // number of local communicators
MPI_Group *mpi_local_group;                 // an array to store local groups
MPI_Comm  *mpi_local_comm;                  // an array to store local comm.
mpi_local_group = (MPI_Group*) malloc(numLocalComm*sizeof(MPI_Group));
mpi_local_comm  = (MPI_Comm*)  malloc(numLocalComm*sizeof(MPI_Comm));

for (int i=0; i<numLocalComm; i++)    // i: counter for local communicators
{
   // For each walker in local communicator (group) i, calculate the global ID
   // (its rank in MPI_COMM_WORLD) and put them in the 'ranks' array
   for (int j=0; j<2*multiple; j++)   // j: counter for walkers in a local comm.
      ranks[j] = i*multiple+j;

   // Create local group
   MPI_Group_incl(world, 2*multiple, ranks, &mpi_local_group[i]);

   // Create communicator for that group
   MPI_Comm_create(MPI_COMM_WORLD, mpi_local_group[i], &mpi_local_comm[i]);
}

free(ranks);

// Now get my local ID in the local communicators
if (myid < (numprocs-multiple))       // Every processor except those in the
{                                     // last window
   comm_id = 2*(myid/(2*multiple));
   MPI_Comm_rank(mpi_local_comm[comm_id], &mylocalid[0]);
   // I am now part of communicator comm_id with local_id[0]=mylocalid[0]
}
else
   mylocalid[0] = INT_MAX;            // just give it an invalid value

if (myid >= multiple)                 // Every processor except those in the
{                                     // first window
   comm_id = 2*((myid-multiple)/(2*multiple)) + 1;
   MPI_Comm_rank(mpi_local_comm[comm_id], &mylocalid[1]);
   // I am now part of communicator comm_id with local_id[1]=mylocalid[1]
}
else
   mylocalid[1] = INT_MAX;            // just give it an invalid value
```

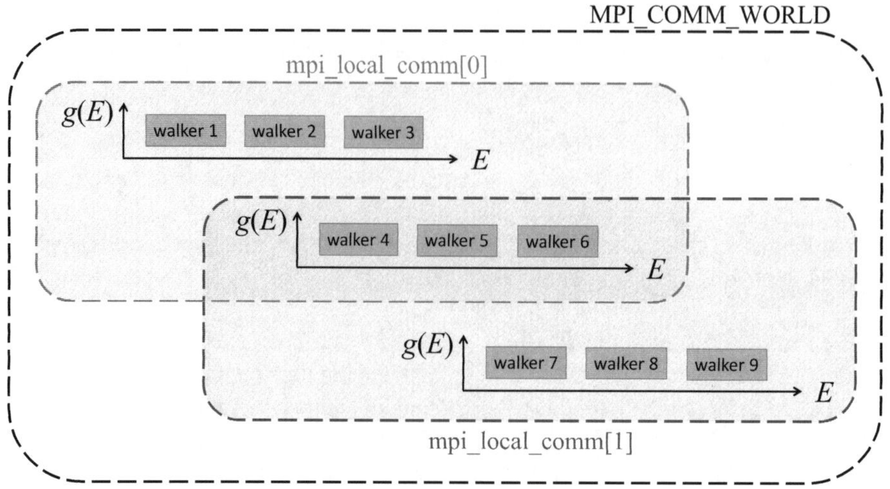

Figure 3. A figure to demonstrate how local communicators are set up.

3.3.2. Exchanging configurations This is the heart of REWL or any replica-exchange simulation in general. There are three main steps:

(i) Find an exchange partner for each walker from the neighboring energy window. These are randomly assigned by one walker in the energy window who then communicates the exchange partner IDs via the `MPI_Scatter` function.

(ii) Each of the assigned pairs then determine the exchange probability following Equation (1). This itself is a three-step process if we just count the MPI communications involved.

  (a) First, both walkers must communicate their current energies to their partner. The function `MPI_Sendrecv_replace` can conveniently be used for this purpose: it exchanges the content of the same variable in both partners. Both partner can then individually calculate their part of contribution ($g_i(E_X)/g_i(E_Y)$; $g_j(E_Y)/g_j(E_X)$ for the other partner) to the total exchange probability. This cannot be done by only one partner because the information about the other's $g(E)$ is not available.

  (b) One of the partners then sends its result via the basic `MPI_Send` function to the other, who uses `MPI_Recv` to receive it, calculates the total exchange probability $P(X \leftrightarrow Y)$, and decides whether an exchange will take place or not by comparing it to a random number.

  (c) This decision is communicated to the partner via the basic `MPI_Send` and `MPI_Recv` functions.

(iii) Finally, the actual configurations are exchanged, if the decision was made to do so. The function `MPI_Sendrecv_replace` is used again.

The following example code demonstrates how this could be implemented. Note how all data exchanges only use local communicators.

```
// Arguments passing into the function:
// - swap_direction: indicates exchange with left or right neighbor window
// - index_akt:      the own, current histogram index
```

IOP Conf. Series: Journal of Physics: Conf. Series **1012** (2018) 012003 doi:10.1088/1742-6596/1012/1/012003

```
int replica_exchange(int swap_direction, int index_akt)
{
   int i_new;              // histogram index of my configuration
   double myfrac;          // my local exchange probability
   double otherfrac;       // local exchange probability of my swap partner
   double wk;              // the combined exchange probability
   double randx;           // a random number between 0 and 1

   int change = 0;         // boolean: 0 for not exchanging, 1 for exchanging
   int swap_partner = -1;  // ID of swap partner (for receive_buffer)

   // +++ Step (i): everyone has to find its swap-partner

   int *pairs;                                   // array to store exchange
   pairs = (int*) malloc(2*multiple*sizeof(int));  // partner for each process

   if (mylocalid[swap_direction] == 0)      // I am the temporary 'master' for
   {                                        // this task
      ...
      // Here, the master randomly assigns pairs of walkers for the exchange,
      // and store this information in the array 'pairs'
   }

   // At this point swap partner assignments get communicated from the
   // master's 'pairs' array to everyone's 'swap_partner' variable
   if ((swap_direction==0) && (myid<(numprocs-multiple)))
     // the walkers from the last window should not swap
   {
      // determine which local communicator to use
      comm_id = 2*(myid/(2*multiple));     // all integer, the '/' is a division
      MPI_Scatter(pairs, 1, MPI_INT, &swap_partner, 1, MPI_INT, 0, \
                  mpi_local_comm[comm_id]);
   }

   if ((swap_direction==1) && (myid>=multiple))   // the walkers from the first
   {                                              // window should not swap
      // determine which local communicator to use
      comm_id = ((myid-multiple)/(2*multiple))*2 + 1;
      MPI_Scatter(pairs, 1, MPI_INT, &swap_partner, 1, MPI_INT, 0, \
                  mpi_local_comm[comm_id]);
   }

   free(pairs);

   if (swap_partner != -1)    // if there is a swap-partner for me...
                              // (if I am at a global boundary, I might not have
                              // a swap partner this time)
   {
      // +++ Step (ii)(a): get histogram index from my swap partner
      // and calculate my contribution to the acceptance probability
```

```
        i_new = index_akt;
        MPI_Sendrecv_replace(&i_new, 1, MPI_INT, swap_partner, 1, swap_partner, \
                             1, mpi_local_comm[comm_id], &status);

        if ((i_new>Emaxindex) || (i_new<Eminindex))
           // received index is out of my energy range
           myfrac = -1.0;
        else
           // calculate my part of the exchange probability (g(myE)/g(otherE))
           myfrac = exp(lngE[index_akt] - lngE[i_new]);

        if (mylocalid[swap_direction] < multiple)
           // I receive the other portion of probability,
           // calculate the total prob., then send the decision to my partner
        {
           // +++ Step (ii)(b): get my partner's part of the exchange probability
           MPI_Recv(&otherfrac, 1, MPI_DOUBLE, swap_partner, 2, \
                    mpi_local_comm[comm_id], &status);

           // calculate combined exchange probability and roll the dice
           if ((myfrac>0.0) && (otherfrac>0.0))
           {
              randx = (1.0*rand()/(RAND_MAX+1.0));
              wk = myfrac*otherfrac;
              if (randx < wk) change = 1; // decision to exchange
           }

           // +++ Step (ii)(c): tell my exchange partner the exchange decision
           MPI_Send(&change,1,MPI_INT,swap_partner,3,mpi_local_comm[comm_id]);
        }
        else     // I send my part of exchange probability and await decision
        {
           // +++ Step (ii)(b): send my exchange probability part to my partner
           MPI_Send(&myfrac, 1, MPI_DOUBLE, swap_partner, 2, \
                    mpi_local_comm[comm_id]);

           // +++ Step (ii)(c): receive exchange decision
           MPI_Recv(&change, 1, MPI_INT, swap_partner, 3, \
                    mpi_local_comm[comm_id], &status);
        }

        // +++ Step (iii): if decision was positive, exchange replicas
        if (change == 1)
           MPI_Sendrecv_replace(&latticepoint[0], numberspins+2+1, MPI_INT, \
           swap_partner, 1, swap_partner, 1, mpi_local_comm[comm_id], &status);
     }

     // Returns whether or not configs were actually exchanged
     return(change);
  }
```

3.3.3. Checking flatness The existing function to check the flatness criterion for a single-walker Wang–Landau program can be extended by the following code to check the flatness of all *other* walkers in the same energy window. At this moment, the integer variable `myflat` has the value 1 if the histogram for the individual walker was determined to be flat, or 0 otherwise.

```
int otherflat;              // a Boolean to store whether other walkers have a
                            // flat histogram or not (0: not flat ; 1: flat)

if (myid%multiple == 0)  // this process serves as the 'master' in the energy
                            // window and receives individual results from all
                            // other walkers
{
   for (int i=1; i<multiple; i++)
   {
      MPI_Recv(&otherflat, 1, MPI_INT, myid+i, 66, MPI_COMM_WORLD, &status);
      myflat *= otherflat;            // if one walker has a '0' (histogram not
                                      // flat), overall result will be '0'
   }

   for (int i=1; i<multiple; i++)    // now let everybody else know
      MPI_Send(&myflat, 1, MPI_INT, myid+i, 88, MPI_COMM_WORLD);
}
else      // others send my flatness to 'master' and receive combined result
{
   MPI_Send(&myflat, 1, MPI_INT, myid-(myid%multiple), 66, MPI_COMM_WORLD);
   MPI_Recv(&otherflat, 1, MPI_INT, myid-(myid%multiple), 88, MPI_COMM_WORLD, \
            &status);

   // replace individual flatness result by combined result
   myflat = otherflat;
}

return(myflat);
```

Note a couple of things here. At the end, the variable `myflat` contains the combined result for all energy windows. That is, the main program for each walker will only call the function to go to the next Wang–Landau iteration if all individual walkers have a flat histogram. In this example we used only the basic `MPI_Send` and `MPI_Recv` functions for demonstration purposes. (See how the two separate messages that every process exchanges with the temporary master process have different tags, 66 and 88). If we would also have introduced local communicators for the walkers within every energy window, as shown in Section 3.3.1, the function `MPI_Allreduce` could have accomplished exactly the same within each local communicator, where the product of the same variable from all processors would be calculated and then broadcast within that local communicator – the value of that variable would be replaced by the product, and every processor would have the same value at the end.

3.3.4. Merging densities of states from different walkers and terminating the simulation If it is determined above that all histograms with the same energy window are flat, all walkers in that energy window would then proceed to the next WL iteration and merge (average) their density-of-states estimators. Without showing the explicit code snippet here, we recommend using the function `MPI_Allreduce` on the local communicators again. The function collects a data array (containing the density of states) from each process in the communicator, reduces the data by applying a predefined reduction function, and then redistributes the result to all processes. The reduction function used here is `MPI_SUM`. The summed density of states can then be divided locally by the number of processes in the communicator to calculate the average. Finally, to decide whether to terminate the simulation, we invoke another `MPI_Allreduce` on `MPI_COMM_WORLD` using the reduction function `MPI_MAX`:

```
MPI_Allreduce(&lnf, &lnf_slowest, 1, MPI_DOUBLE, MPI_MAX, MPI_COMM_WORLD);
```

to check whether any of the modification factors f_i from the individual energy window is still above the predefined termination value, in which case the simulation simply keeps running.

4. REWL: FAQ

Over the past few years we received many e-mails and questions during workshops asking for practical advice on certain implementation details of REWL. Most of these questions are recurring and of general relevance, so we would like to answer them in the following.

List of topics:

4.1) Parallel programming model: OpenMP vs. MPI
4.2) Alternatives for histograms with fixed bin size
4.3) Convergence of REWL with equal-size energy windows
4.4) Why is replica-exchange necessary at all?
4.5) Creation of initial configurations
4.6) What if new low-energy states are found?
4.7) What to do with rejections?
4.8) End of simulation
4.9) Miscellaneous questions

4.1. Parallel programming model: OpenMP vs. MPI

Q: *Between OpenMP [27] and MPI, which one is more efficient / effective to write the parallel replica-exchange Wang–Landau code?*

A: We recommend using MPI because it will work on machines with both shared and distributed memory. If the code is written using MPI it will be much more portable if you want to run it for larger systems, with more walkers, or on different supercomputers. If you use OpenMP [27] you will be restricted to shared-memory systems, typically a single node on a supercomputer. At that point the efficiency is of secondary importance.

4.2. Alternatives for histograms with fixed bin size

Q: *For various reasons it is difficult to attain a flat histogram for systems with a continuous energy function. Choosing a good bin size is a constant puzzle/inconvenience/annoyance. Are there ways to solve this problem?*

A1: Discretization (or binning) of the energy range for a system with a continuous energy function often introduces systematic errors; extra care has to be taken to choose a suitable bin size. One possible solution is to use the so-called kernel or global updates, where multiple histogram bins are simultaneously updated by different amounts according to a kernel function. This allows, in principle, for the energy bins to be arbitrarily narrow. Different kernel functions have been proposed and tested [28]. Gauss kernels are very popular to update binned thermodynamic functions in different WL-type methods in the molecular-dynamics community [29]. It was recently shown that simulations can be significantly sped up when using global updates even when using histograms with bin sizes about ten times smaller compared to simulations with single-bin updates [29].

A2: Another workaround is to avoid the use of a histogram and binning, in general. Recently, a histogram-free version of multicanonical / Wang–Landau sampling has been proposed to address this problem [30]. The energies generated during the simulation are stored directly in a data set instead of in a histogram, which allows for the global update of the estimation of the density of states. Although the sampling weight is still adjusted based on the (inverse of the) estimator of the density of states, this algorithm employs a statistical test as the criterion to proceed to the next iteration instead of a flat histogram. The use of a histogram is thus not necessary.

4.3. Convergence of REWL with equal-size energy windows

Q: *Every walker proposes a replica-exchange every N steps of ordinary Wang–Landau random walk. The parameter N is universal among walkers, so the implementation effectively synchronizes every N steps. Essentially, all walkers will perform exactly (or almost exactly) the same number of random walks. Therefore, the underlying area of the histogram will be the same. However, as the distribution $g(E)$ is intrinsically different in every energy window (therefore the areas under the $g(E)$ curve are different), if we ensure the same number of walks in each window, the histogram [and $g(E)$] in some windows will be consistently above other windows. I think this is a problem for the Wang–Landau algorithm to converge. How do you work around it?*

A: Undoubtedly, N is universal and a way to intrinsically synchronize all walkers with the assumption that all walkers run equally fast, which is reasonable since they all do the same amount of operations. So it should not slow the whole procedure down significantly. Note that when reducing the modification factor f, we only wait until all walkers with the *same* window fulfill the flatness criterion. Walkers in different windows can be in different iterations of the WL scheme (that is, they have different f) but all walkers with the same window always have the same f. This is not a problem as the replica-exchange is still valid. But if the walkers in the same window had different f, that would be a problem.

While there is no convergence problem in principle, it is true that this is far from optimal. In the optimal case the number and widths of your energy windows should be chosen in a way that all walkers converge at about the same time. That means, each energy window has *different* numbers of energy bins. See Figure 1b in [1], for example. Of course, in principle, this is impossible since one does not know the density of states and the amount of work needed to estimate it beforehand. We did the run for the mentioned Figure 1b after we already knew the solution to the problem just to demonstrate the effect of optimized energy windows on the efficiency. In practice, one could use the information gathered in pre-runs (see below) to estimate a good bin number for different windows. However, investigations have also shown that the convergence behaviors for different energy windows could change and desynchronize towards the end of the simulation [31]. This is an unsolved question and is a research problem itself.

4.4. Why is replica-exchange necessary at all?

Q: *If we perform independent WL runs in each window without the replica-exchange step, will the algorithm still produce reasonably good results? I recall that the error mostly came from the boundary of the window, and since we have overlapping windows, it may not be a problem.*

A: This was actually done in an early attempt to parallelize WL [18]. While this can be beneficial and lead to the correct result for some systems, it prevents the algorithm, in general, from reaching all regions of the phase space, in particular those associated with low temperatures. In fact, there have been ergodicity problems with certain "difficult" systems such as those with first-order transitions. An example is the study of surface adsorption of the hydrophobic-polar lattice model proteins [32]. Some low-energy states, which are known to exist, were never generated because the random walker in the lowest energy window would never "heat up" enough to overcome the entropic barrier around that state. In other words, the walker could be trapped in a local minimum in the energy landscape and would only be able to escape this by exchanging configurations with a higher-energy ("hotter") window.

4.5. Creation of initial configurations

Q: *At the beginning of the simulation, how to find the starting configuration / energy for each energy window? Do I just randomly create configurations until, eventually, I created at least one for every energy window? Can I just copy over configurations from one window into another if they are in the overlap regions between two windows?*

A: As you say, one way is for each energy window to start from a randomly chosen configuration, which probably has an energy that is outside of the energy window. Then you could apply Monte Carlo moves to generate new configurations, until an energy that falls within the energy window is discovered. However, this can take an arbitrarily long time for large systems and is therefore probably not the best way of doing it.

For systems with known ground states, for example the Ising model, one can start with that ground state configuration and perform a Metropolis simulation at high temperatures (or even infinite temperature, for that matter, just accept every trial move). Whenever a configuration with an energy $E_{i,\mathrm{c}} \pm \epsilon$ is created (where $E_{i,\mathrm{c}}$ is the energy at the center of energy window i and ϵ a reasonable energy range), it can be taken as the initial configuration for energy window i. This method is very fast and was used to create initial configurations for the spin-model simulations shown in [1, 2]. It took only a few minutes to create initial configurations even for the largest system sizes that have been studied.

Another way is to perform a WL pre-run on the global energy range and store the configurations that can be used as the starting configurations for each energy windows in a similar way. This way might be slower, but could be the only way if low-energy states are unknown and a pre-run is necessary for determining the relevant energy range anyway (see question 4.6). In a similar manner, initial states can be created "along the way" if a pre-run is performed to determine suitable energy window ranges for runtime balancing, see question 4.3.

4.6. What if new low-energy states are found?

Q: *For a model of which the ground state energy and the energy landscape are not known a priori, what is the best way to start: performing parallel-tempering pre-runs to estimate the global energy range, or starting right away with REWL and adapting energy boundaries on the fly?*

The first would cost much more computing time, the latter might be problematic with regard to convergence.

A: When simulating a new model, most of the time we are in that situation where the ground state information is missing. Both suggested ways of dealing with this problem are possible and are used in practice. Typically, we prefer performing pre-runs to estimate the ground state energy. One advantage is that during these pre-runs we can also generate initial configurations for different energy windows (see also question 4.5). These pre-runs have been very fast for most of our studies, in particular because it is often sufficient to know *a few* very-low energy states, rather than *the* lowest. For example, if we are interested in studying a freezing transition, it might be sufficient to know a couple of states from the crystalline phase to determine the energy ranges. An alternative to parallel-tempering pre-runs could be conventional WL runs with an excessively large energy range for which the histogram never becomes flat.

But sometimes, the sampling of the actual ground states is in fact the interest. In such cases it is inevitable to adapt energy ranges and energy windows every time a new lowest energy is found. It is important, however, to restart the whole WL scheme after making this change. That is, all histograms need to be deleted and f reset to the initial value. There have been attempts to avoid a complete restart [33]. While this is an effective approach to aid the discovery of the ground states, it is harder to control the behavior of the simulation and could potentially result in artificial signals in thermodynamic functions.

4.7. What to do with rejections?

Q: *While performing the ordinary WL random walk there could be configurations generated that fall outside of the energy window. Do you still increase the histogram entry for the configuration after such a trial move?*

A: Yes, that should always be done in any WL simulation. You certainly reject the newly proposed state which is outside the energy range, but very importantly, count the original energy again. That is, update the corresponding density of states and histogram entries. See [34].

Q: *When proposing a replica-exchange, it could be that E_X is not in its target energy window or E_Y is not in the home energy window. While the replica-exchange is rejected, do you update $H(E)$?*

A: Yes, and also $g(E)$, for the original energy. This situation is, in fact, the same as the one described directly above. Alternatively, this could be interpreted as a Monte Carlo move with zero acceptance probability (see question directly below).

Q: *If the replica-exchange is rejected by chance according to the acceptance probability, do you update $H(E)$?*

A: Yes, and also $g(E)$, for the original energy. This is always done in any Monte Carlo scheme.

4.8. End of simulation

Q: *If the walkers in every energy window can proceed to the next WL iteration on their own pace, some walkers will reach their final modification factor and terminate the WL sampling before others. How to assure that a replica-exchange is only proposed with "living" walkers, that have not terminated the simulation yet?*

A: We suggest doing it differently: keep all walkers alive until all others reach the termination criterion as well. This is for exactly the reason of assuring that there will always be a replica-exchange partner for every walker. In our implementation, walkers who have finished their WL iterations just keep walking, without updating f anymore, to be able to provide exchange configurations. All walkers terminate simultaneously when all have reached the termination criterion individually. See section 3.3.4 for an implementation example. It would be optimal if all walkers would reach the termination criterion naturally at the same time. For this to happen one would have to create energy windows of different sizes. We have experimented with this approach and shown an example in [2], Figure 1c. The total runtime for each window is, again, estimated based on measurements from pre-runs.

4.9. Miscellaneous questions

Q: *When there are multiple walkers with the same energy window, a walker may receive more than one proposal for replica-exchange from its neighboring walkers. Do you only accept the first and ignore all others? How do you handle this situation?*

A: For the replica-exchange, your scheme seems valid but might not be optimal. We randomly draw *pairs* of walkers from neighboring windows. So, by construction, no walker can receive more than one proposal at a time. This also ensures that every walker gets a proposal in the first place, assuming there are equally many walkers in each of the neighboring energy windows.

Q: *How do you choose N, the number of ordinary Monte Carlo sweeps before a replica-exchange step is proposed? Any typical choice? How does it affect the results?*

A: Unfortunately, that mainly depends on the physical system at hand. Optimally, you want to give the walkers enough time to "diffuse" through the energy window between replica-exchanges. That is, give them enough time to be able to "randomly walk to the other side" of the energy window. Ideally, measurements are made in pre-runs and N is adapted accordingly before the final run. A bad choice of N would typically affect the performance much more than the actual results in the sense that it can take much longer to converge to the correct results.

Q: *What data are actually exchanged during the replica-exchange step?*

A: The walkers from neighboring energy windows *only* exchange their current configurations (including the corresponding energies). Any other data, in particular $g(E)$ and $H(E)$, remain with the walkers. From the point of view of each of the two walkers individually, this can be seen as "randomly" proposing a new configuration for the next Monte Carlo step.

Q: *It is mentioned that "$g(E)$ from all m walkers in the same energy window are averaged and redistributed". What do you mean by "redistributed"? Do you try to "stitch" the $g(E)$ among energy windows? (bring them into the same height?)*

A: At the end of an iteration, the $g(E)$'s from all walkers in that window are averaged, and then every walker in that window proceeds with that average. That is, after the averaging, all walkers in the energy window have the same density of states to start the new iteration. There is no need to join the density-of-states estimators at this point.

Acknowledgments
We thank N. Casper, Y. Ren, Y. Shevchenko, T. K. Bose, and D. Moreira for lively discussions and acknowledge their contributions with regard to contents of the FAQ section of this paper. Y.W.L. was sponsored by the Oak Ridge Leadership Computing Facility at the Oak Ridge National Laboratory, which is supported by the Office of Science of the U.S. Department of Energy under Contract No. DE-AC05-00OR22725.

References

[1] Vogel T, Li Y W, Wüst T and Landau D P 2013 *Phys. Rev. Lett.* **110** 210603
[2] Vogel T, Li Y W, Wüst T and Landau D P 2014 *Phys. Rev. E* **90** 023302
[3] Vogel T, Li Y W, Wüst T and Landau D P 2014 *J. Phys.: Conf. Ser.* **487** 012001
[4] Gai L, Vogel T, Maerzke K A, Iacovella C R, Landau D P, Cummings P T and McCabe C 2013 *J. Chem. Phys.* **139** 054505
[5] Li Y W, Vogel T, Wüst T and Landau D P 2014 *J. Phys.: Conf. Ser.* **510** 012012
[6] Perera D, Vogel T and Landau D P 2016 *Phys. Rev. E* **94** 043308
[7] Valentim A, Rocha J C S, Tsai S H, Li Y W, Eisenbach M, Fiore C E and Landau D P 2015 *J. Phys.: Conf. Ser.* **640** 012006
[8] Shevchenko Y A, Makarov A G, Andryushchenko P D and Nefedev K V 2017 *J. Exp. Theor. Phys.* **124** 982
[9] Ren Y 2016 private communication. Work lead to the development of another parallelization scheme for Wang–Landay simulations [10]
[10] Ren Y, Eubank S and Nath M 2016 *Phys. Rev. E* **94** 042125
[11] Casper N 2016 *Stochastic analytic continuation of Quantum Monte Carlo data* Master's thesis Technical University Braunschweig, Germany
[12] Yuk S F, Li Y W, Cooper V R and Eisenbach M 2017 private communication
[13] Wang F and Landau D P 2001 *Phys. Rev. Lett.* **86** 2050
[14] Wang F and Landau D P 2001 *Phys. Rev. E* **64** 056101
[15] Geyer C J 1991 Markov chain Monte Carlo maximum likelihood *Computing Science and Statistics, Proceedings of the 23rd Symposium on the Interface* (Interface Foundation of North America) p 156 retrieved from the University of Minnesota Digital Conservancy URL `http://hdl.handle.net/11299/58440`
[16] Hukushima K and Nemoto K 1996 *J. Phys. Soc. Japan* **65** 1604
[17] Demonstration codes complementary to the Replica-Exchange Wang–Landau sampling tutorial lectures given at the IX Brazilian Meeting on Simulational Physics, August 21–25, 2017, Natal, Brazil. `https://github.com/yingwaili/bmsp2017`
[18] Cunha-Netto A G, Caparica A A, Tsai S H, Dickman R and Landau D P 2008 *Phys. Rev. E* **78** 055701
[19] MPICH – High-Performance Portable MPI. `http://www.mpich.org/`
[20] MVAPICH: MPI over InfiniBand, Omni-Path, Ethernet/iWARP, and RoCE. `http://mvapich.cse.ohio-state.edu/`
[21] Open MPI: Open Source High Performance Computing. `https://www.open-mpi.org/`
[22] Lawrence Livermore National Laboratory's MPI tutorial. `https://computing.llnl.gov/tutorials/mpi/`
[23] Message Passing Interface (MPI) Forum - contains the official documents for the MPI Standards. `http://www.mpi-forum.org/`
[24] Pacheco P S 1996 *Parallel Programming with MPI* (San Francisco, CA, USA: Morgan Kaufmann Publishers Inc.) ISBN 1-55860-339-5
[25] Bland W, Bouteiller A, Herault T, Bosilca G and Dongarra J 2013 *Int. J. High Perform. Comput. Appl.* **27** 244
[26] Bland W, Du P, Bouteiller A, Herault T, Bosilca G and Dongarra J 2013 *Concurr. Comput. Pract. Exper.* **25** 2381
[27] OpenMP - The OpenMP API specification for parallel programming. `http://www.openmp.org/`
[28] Zhou C, Schulthess T C, Torbrügge S and Landau D P 2006 *Phys. Rev. Lett.* **96** 120201
[29] Junghans C, Perez D and Vogel T 2014 *J. Chem. Theory Comput.* **10** 1843
[30] Li Y W and Eisenbach M 2017 A Histogram-Free Multicanonical Monte Carlo Algorithm for the Basis Expansion of Density of States *Proceedings of the Platform for Advanced Scientific Computing Conference* PASC '17 (New York, NY, USA: ACM) p 10:1
[31] Zhao Y, Cheung S W, Li Y W and Eisenbach M 2014 *Physics Procedia* **57** 43
[32] Li Y W, Wüst T and Landau D P 2013 *Phys. Rev. E* **87** 012706
[33] Wüst T and Landau D P 2008 *Computer Physics Communications* **179** 124
[34] Schulz B J, Binder K, Müller M and Landau D P 2003 *Phys. Rev. E* **67** 067102

IOP Conf. Series: Journal of Physics: Conf. Series **1012** (2018) 012004 doi:10.1088/1742-6596/1012/1/012004

Ising Criticality of the Clock Model from Density of States Obtained by the Replica Exchange-Wang-Landau Method

Antonio Cadilhe

Instituto de Física, Universidade Federal da Bahia, 40210-340, Salvador, Brazil

Abstract. We performed extensive simulations, using the Replica Exchange-Wang-Landau method, of the clock model for orders 3 and 4 on a square lattice, where critical behaviors are expected to belong to the Ising universality class. Though order 2 represents the Ising model, thus, being exactly solvable in two-dimensions, we still provide such results for comparison to the other two orders. Results for various energy related quantities such as the mean energy per spin, specific heat, as well as logarithm scaling of the peak of the specific heat are presented and shown to follow Ising behavior. Additionally, we also present results related to magnetic quantities, such as the magnetization, magnetic susceptibility, and corresponding scaling behavior of the peak of the magnetic susceptibility. Again, our results show scaling in conformity to Ising critical behavior.

1. Introduction

At the essence of the Wang-Landau method, [1, 2, 3] a random walk in energy space is performed, while building the density of states of a particular system. Concomitantly, an histogram of visited energy values is also built providing a way to warrant its flatness at some specified level of quality. Once the flatness of the histogram has been signaled, the histogram is reset and the simulation proceeds to the next iteration level using a smaller factor for the buildup of the density of states, until a pre-defined, small enough value is reached.

Since its debut, the Wang-Landau method has been shown to consistently produce high-quality density of states of several model systems, such as lattice spin and biological systems. [1, 2] In its present version, the Replica Exchange-Wang-Landau (RE-WL), the method builds the density of states without little or prior knowledge of the physical properties of the system, under a set of, well-defined, quality factors. In most systems, a simulation can be carried out by selecting an overall *energy window* and splitting it into several smaller, equally sized, *energy windows*, but in some systems additional knowledge of the *proper* window sizes for energy windows near the energy of the fundamental state(s) of the system may become a requirement.

One such class of possible applications of the RE-WL method is provided by (magnetic) lattice spin systems of wide interest in statistical physics. These systems, typically exhibit critical behavior, in the thermodynamic limit, which the method appropriately addresses. The critical behavior may include well-defined temperatures of first and second order phase transitions or a critical line of temperatures as in the case of the Berenzynski-Kosterlitz-Thouless transition. [4]

An example of such a system, it is provided by the clock model, where, depending on its order, the model can exhibit either Ising or BKT behavior at criticality. As the name implies,

IOP Conf. Series: Journal of Physics: Conf. Series **1012** (2018) 012004 doi:10.1088/1742-6596/1012/1/012004

an order n, clock model proposed by Cyril Domb in the early fifties of the last century, [5, 6] possible, equally-distributed spin orientations are provided by the *hand of a clock*. Consequently, the possible states of a lattice spin can be expressed as

$$\sigma = \left(\cos\left(2\pi\frac{s}{n}\right), \sin\left(2\pi\frac{s}{n}\right)\right), \quad (1)$$

where s takes values 0, 1, ..., $n-1$. The Hamiltonian is of the form

$$H = -J\sum_{\langle i,j\rangle} \sigma_i \cdot \sigma_j, \quad (2)$$

where $J > 0$ is a constant representing the strength of the interaction, σ_i represents a spin positioned at lattice site i, $\sigma_i \cdot \sigma_j$ represents the scalar product between spins i and j, and $\langle i, j\rangle$ denotes sum over nearest-neighbor spin pairs. In this work, we provide initial results of the clock model for orders 3 and 4, i.e., of orders within the Ising universality class. [7, 8, 9] Results for order 2, the Ising model, are also included to provide direct comparison to the other two orders.

2. Clock Model

As above-mentioned, depending on the order of the clock, its critical behavior can change from the Ising class to the BKT class. [10] Within the scope of the present work, we restrict ourselves to the Ising class, which implies clock orders of 2 to 4. Orders $n \geq 5$ are within the BKT universality class and, therefore, outside the scope of the present work. The critical temperatures for $n = 2$, 3, and 4 were exactly derived, by renormalization group studies, [10] to be, respectively, $2/\ln(1+\sqrt{2})$, $3/(2\ln(1+\sqrt{3}))$, and $1/\ln(1+\sqrt{2})$. Note that the critical temperature for $n = 4$ is just half of the Ising model, as it represents two mutually orthogonal Ising models.

Let us, now, consider a square lattice, where spins σ_i lie at each of its sites and the subscript labels i-th site of the lattice. In the case of the clock model of order n, each spin can assume n possible states (orientations) given by Eq. (1) and the Hamiltonian for the clock model of order n is defined by Eq. (2).

Finally, it is pedagogical to note that in the limit $n \to \infty$, the clock model is expected to behave as the planar rotator model, in which, each spin can continuously vary its orientation, but we will not address here such case.

3. Simulations and Results

Extensive Monte Carlo simulations, using the RE-WL method of the clock model within an overall energy window, per spin, of $[-2, 0]$ were performed for orders 2, 3, and 4 with a flatness criterium for the minimum value of the histogram set to 70% of its average value and a stop criterium set to $\ln(f) = 10^{-8}$ [1].

Each of the clock orders above were simulated for system sizes of $L = 20$, 40, 80, 100, and 200 lattice constants with a number of samples of 10, 2, 2, 2, and 1, respectively. The number of energy windows varied with the system size being 4, 4, 10, 20, 40, respectively, for the previous sequence of system sizes. In all cases, the overall energy window of each simulation per spin for the various orders of the clock model, i.e., the physical energy interval per spin, was set to be $[-2, 0]$, which provides ample energy variation, allowing proper description of various quantities up to including the fundamental state along with its critical behavior.

We start presenting our results by providing the results of a sample of the fundamental quantity $g(e)$, the density of states, in Fig. 1 for a system with 100×100 spins on a square

[1] Please refer to references [1, 2, 3] for detailed descriptions of the Wang-Landau method and clarification on the meaning of the parameters.

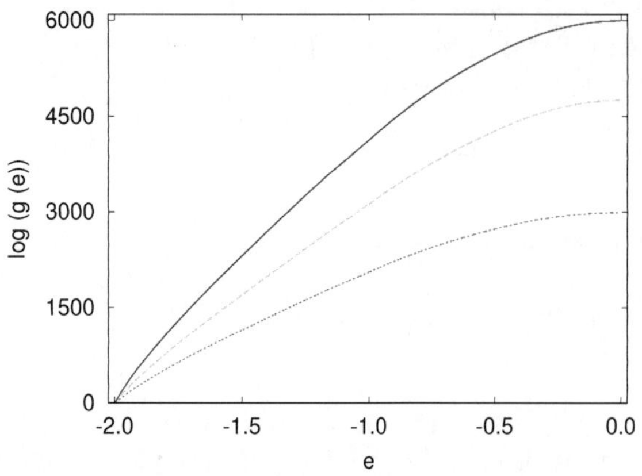

Figure 1. Plots of the (decimal) logarithm of the density of states, $\log g(e)$, of orders 2 (red), 3 (green), and 4 (blue) of systems with 100×100 spins. Error bars (not provided) are smaller than line thickness of the plots.

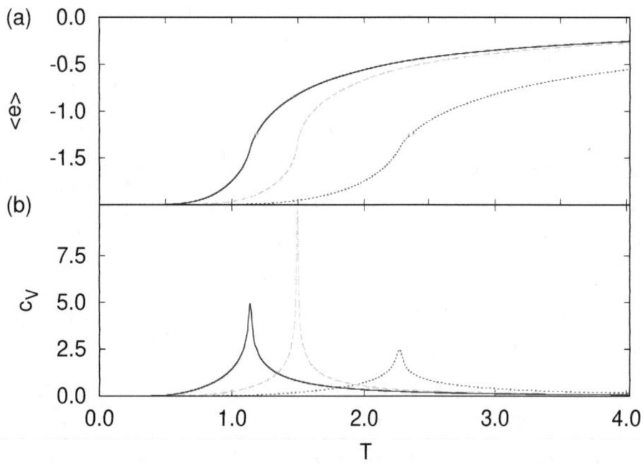

Figure 2. Plots of quantities associated with the energy for the three clock orders 2 (red), 3 (green), and 4 (blue). In parts (a) and (b) the plots are for a square lattice of 100×100 spins. (a) Average energy per spin, $\langle e \rangle$, as a function of temperature. (b) Plots of the specific heat c_V.

lattice. We note that the error bars associated are smaller than the thickness of the lines for this and remaining (not shown) system sizes. In the following, the Ising model will be solely included for comparison purposes, since the model is exactly solvable in two dimensions. The (logarithm of the) density of states grows substantially with the order of the clock model. In fact, near the maximum value of $g(e)$, the values are about $2.88 \cdot 10^3$, $4.76 \cdot 10^3$, and $6 \cdot 10^3$ orders of magnitude greater than $g(-2)$, i.e., the density of states at energy of the fundamental state. Additionally, order 3, at the maximum has a factor of about 1.65 greater of the number of orders of magnitude of the corresponding value of the Ising model, while order 4 the number of orders of magnitude is just 1.26 times larger of those of order 3.

The mean energy per spin is defined $\langle e \rangle = \langle E \rangle / L^2$, where $\langle E \rangle$ is the average energy of the system. The canonical partition function

$$Z(T) = \sum_E g(E) e^{-\beta E}, \tag{3}$$

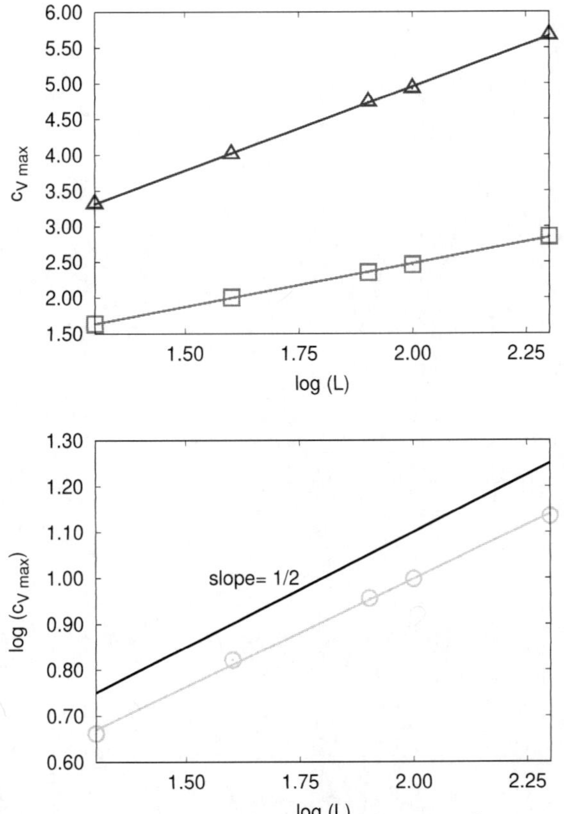

Figure 3. Scaling of the specific heat peaks, $c_{V\mathrm{max}}$, for the three clock orders 2 (red), 3 (green), and 4 (blue). (a) Scaling of the specific heat peaks as a function of the logarithm of the system size for orders 2 and 4. (b) Log-log plot for order 3, where the black solid curve has slope 1/2.

where $\beta = 1/k_BT$, and k_B is the Boltzmann constant, can be utilized to define the mean energy, per spin,

$$\langle e \rangle = Z^{-1} \sum_E \frac{E}{N} e^{-\beta E}. \tag{4}$$

Next, we present in Fig. 2(a) the mean energy per spin over a broad interval of temperatures, T, in the range $]0, 8]$. Notice that the Ising model does not merge its mean energy values, per spin, with those of the other two cases, even at the highest temperature of the interval. In contrast, it is somewhat surprising that reasonably merged mean energy values, for $T > 4$, with solely 3 or 4 states per spin. Once again, error bars are not included as they are smaller than line thicknesses.

Next, we show plots of the specific heat (per spin) defined as

$$c_V = \frac{\langle E^2 \rangle - \langle E \rangle^2}{N k_B T^2}, \tag{5}$$

in Fig. 2(b). Of notice, it is the enhanced c_V of the order 3 clock relatively to the other two orders. The clock orders 2 and 4 also have a logarithmic dependence of the peak of the specific heat with size shown in Fig. 3(a), while order 3 has a power-law dependence as shown in Fig. 3(b).

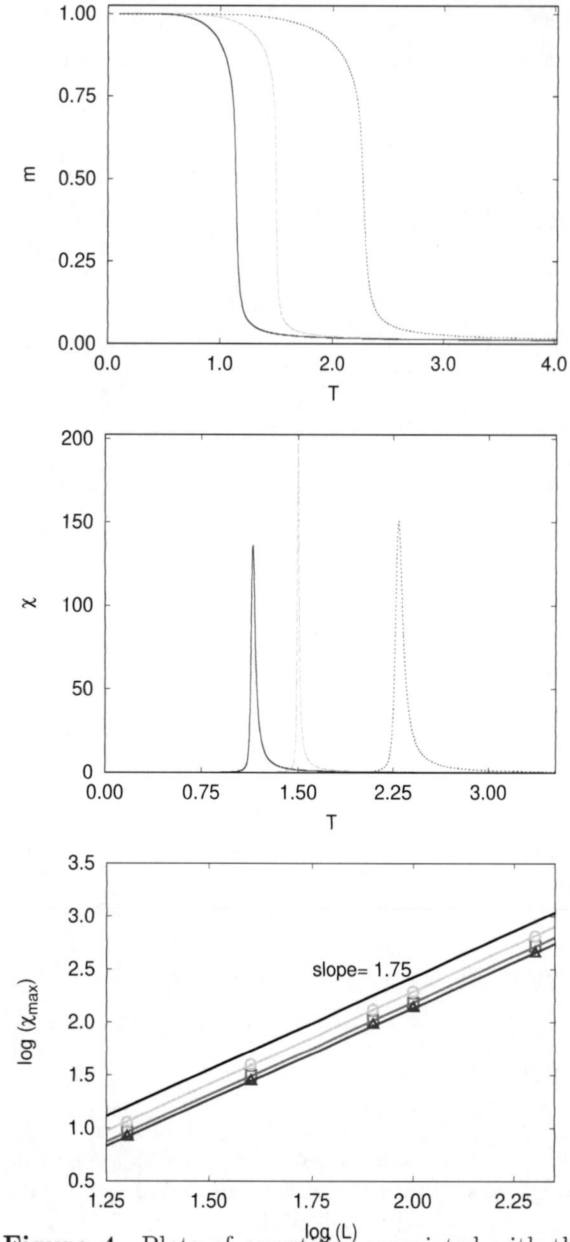

Figure 4. Plots of quantities associated with the magnetization for the three clock orders 2 (red), 3 (green), and 4 (blue). In parts (a) and (b) the plots are for a lattice of 100×100 spins. (a) Magnetization, m, as a function of temperature. (b) Plots of the magnetic susceptibility χ. (c) Log-log plot showing the scaling of the magnetic susceptibility peaks, $\chi_{\max}$, with system size. Solid, black line shows expected slope of $1\frac{3}{4}$ for Ising behavior.

Values of the slopes of orders 2 and 4 are not universal, so their values vary. The computed slope of the log-log plot is 0.47, which differs from the expected 1/3. [6]

Now, we move on to show results of magnetic related quantities. We start with the magnetization, the order parameter, defined as

$$m = \frac{\sqrt{(\sum_{i=1}^{N} \sigma_{xi})^2 + (\sum_{i=1}^{N} \sigma_{yi})^2}}{N}, \tag{6}$$

where $\sigma_{\alpha i}$ represents the corresponding coordinate, given by $\alpha = \{x, y\}$, of the spin at site i. The above definition in the case of orders 2 and 4 boils down to

$$m = \frac{|\sum_{i=1}^{N} \sigma_i|}{N}, \tag{7}$$

for both cases. In the case of the Ising model (order 2) we just need to associate σ_i to σ_{yi} as this is the only non-zero coordinate. Now, order 4 represents two Ising models at orthogonal directions, i.e., the magnetization can be formally written as

$$m = \frac{|\sum_{i=1}^{N} (a_i \sigma_{xi} + (1 - a_i)\sigma_{yi}|}{N}, \tag{8}$$

where a_i takes value 1 (one) if the spin is aligned along the x-direction and 0 (zero) if the spin is aligned along the y-direction. We observe in Fig. 4(a) plots of the magnetization for temperatures in the interval $]0, 4]$. As expected, the increased number of states per spin with the order, leads to the transitioning of the magnetization, from its maximum to approximately zero, occurring at a lower temperatures, i.e., the critical behavior occurs at a lower critical temperature with increasing orders (within the set 2, 3, and 4).

Another quantity of interest is the magnetic susceptibility defined as

$$\chi = N \frac{\langle m^2 \rangle - \langle m \rangle^2}{k_B T}. \tag{9}$$

Plots of the magnetic susceptibility are shown in Fig. 4(b). The results corroborate the higher response, at criticality, of order 3 as observed for c_V. Scaling of the maximum of the magnetic susceptibility follows a dependence of the form

$$\chi \approx L^{\gamma/\nu}. \tag{10}$$

Since the exponent associated with the correlation length, $\nu = 1$, for the Ising model in two dimensions, we are directly observing the value of γ (see references [12, 11]). This is, in fact, observed in Fig. 4(c) where the three curves have the expected slope of 7/4 illustrated by the solid (black) line.

4. Concluding Remarks

We presented preliminary results of the clock model restricted to orders 3 and 4, where its critical behavior belongs to the Ising universality class.

The results fully corroborate expected properties of the Ising universality class, such as the logarithmic divergence of the specific heat (per spin) and the power-law divergence of the magnetic susceptibility. In particular, our results on the critical temperatures, not included in the present work, are in close agreement with exact results from renormalization group studies. [10]

Acknowledgments

The author gratefully acknowledges partial financial support from a FAPEMIG (PACCSS/CEX-30034-12) grant.

References

[1] F. Wang and D.P. Landau, *Efficient, Multiple-Range Random Walk Algorithm to Calculate the Density of States*, Phys. Rev. Lett. **86**, 2050 (2001).

[2] F. Wang and D.P. Landau, *Determining the Density of States for Classical Statistical Models: A Random Walk Algorithm to Produce a Flat Histogram*, Phys. Rev. E**64**, 056101 (2001).

[3] Y.W. Li, T. Vogel, T. Wüst, and D.P. Landau, *A New Paradigm for Petascale Monte Carlo Simulation: Replica Exchange Wang–Landau Sampling*, J. Phys.:Conf. Series **510**, 012012 (2014).

[4] J.V. José (ed.), *40 Years of Berezinskii-Kosterlitz-Thouless Theory*, (World Scientific, Singapore) (2013).

[5] R.B. Potts, *Some Generalized Order-Disorder Transformations*, Proc. Cambridge Philos. Soc. **48**, 106 (1952).

[6] F.Y. Wu, *The Potts Model*, Rev. Mod. Phys. **54**, 235 (1982).

[7] L. Onsager, *Crystal Statistics. I. A Two-Dimensional Model with an Order-Disorder Transition*, Phys. Rev. **65**, 117 (1944).

[8] P.W. Kasteleyn, *Dimer Statistics and Phase Transitions*, J. Math. Phys. **4**, 287 (1963).

[9] L.E. Reichl, *A Modern Course in Statistical Physics*, (John Wiley, New York, USA) (1998).

[10] G. Ortiz, E. Cobanera, and Z. Nussinov, *Berenzinskii-Kosterlitz-Thouless Transition through the Eyes of Duality*, in Ref [4].

[11] H.E. Stanley, *Introduction to Phase Transitions and Critical Phenomena*, (Clarendon, Oxford, UK) (1971).

[12] V. Privman (ed.), *Finite Size Scaling and Numerical Simulation of Statistical Systems*, (World Scientific, Singapore) (1990).

IOP Conf. Series: Journal of Physics: Conf. Series **1012** (2018) 012005 doi:10.1088/1742-6596/1012/1/012005

On the use of the energy probability distribution zeros in the study of phase transitions

L A S Mól[1], R G M Rodrigues[1], R A Stancioli[1], J C S Rocha[2] and B V Costa[1]

[1] Laboratório de Simulação, Departamento de Física, ICEx, Universidade Federal de Minas Gerais, 31720-901, Belo Horizonte, Minas Gerais, Brazil

[2] Departamento de Física, ICEB, Universidade Federal de Ouro Preto, 35400-000, Ouro Preto, Minas Gerais, Brazil

E-mail: lucasmol@fisica.ufmg.br

Abstract. This contribution is devoted to cover some technical aspects related to the use of the recently proposed energy probability distribution zeros in the study of phase transitions. This method is based on the partial knowledge of the partition function zeros and has been shown to be extremely efficient to precisely locate phase transition temperatures. It is based on an iterative method in such a way that the transition temperature can be approached at will. The iterative method will be detailed and some convergence issues that has been observed in its application to the 2D Ising model and to an artificial spin ice model will be shown, together with ways to circumvent them.

1. Introduction

Phase transitions are a core subject in statistical physics. Although the concept of phase itself is somewhat ill-defined, phase transitions are much better stated. Indeed, phases can only be clearly defined when phase boundaries or phase coexistence are present, allowing one to identify the differences between them. Phase transitions, on the other hand, are characterized by a non-analytic behavior of the free energy, which in general leads to divergences and discontinuities in thermodynamic quantities that can be "easily" identified. However, given the interactions among the elementary degrees of freedom of a system, the identification and characterization of possible phase transitions is not an easy task. As is well known, only very few systems can be exactly solved, in such a way that approximated methods are necessary. Among many possibilities, computer simulations based on the Monte Carlo method[1] together with finite size scaling theory[2] can give very accurate results, being thus the preferred method for many studies.

In computer simulations only finite systems can be studied and since phase transitions only exist in the thermodynamic limit, the use of finite size scaling theory to characterize phase transitions is mandatory. In fact, as the free energy is obtained by the natural logarithm of the partition function and the latter is a sum of positive terms, non-analyticities can only appear in the free energy density when the thermodynamic limit is taken. Nevertheless, pseudo-phase transitions are present in finite systems and play a very important role in modern physics. A key example are conformational transitions that occur in macromolecules [3]. Polymers are intrinsically finite systems, specially biological polymers (e.g. proteins) in which a different number or arrangement of monomers imply a completely different biological function. Protein folding, for example, can be thought of as a pseudo-phase transition between unfolded and folded states and its comprehension is of extreme importance. Then, methods that are devoted not only to study phase transitions in thermodynamic limit but also to properly explore pseudo-transitions in finite systems are of key importance.

Recently[4] some of us proposed to use the partial knowledge of the energy probability distribution function to obtain a set of zeros from which phase transitions and pseudo-phase transitions can be identified. Among its advantages, the fact that the transition temperature can be approached at will and that ambiguities related to the location of a pseudo-transition temperature in finite systems are not present are of paramount importance. Actually, when considering pseudo-transitions in finite systems each thermodynamic quantity signals the transition in a slightly different temperature, making its proper location ambiguous. By using the zeros of the energy probability distribution, such ambiguity is removed, leading to a single estimate of the pseudo-transition temperature for a given system. In what follows we briefly describe the method and detail its use and application. Some convergence issues are also shown together with ways to circumvent it. We then finish by presenting our closing remarks

2. Energy probability distribution zeros

As already mentioned, phase transitions are characterized by a non-analytic behavior of the free energy. This behavior can be captured by considering the complex zeros of the partition function. Indeed, Yang and Lee [5] have shown that the partition function zeros in the complex fugacity plane contain all the relevant information about the thermodynamic behavior of a system. In particular they showed that in the thermodynamic limit the density of zeros completely determine its critical behavior. In 1964 Fisher [6] extended their idea to the complex temperature plane (Fisher zeros). In summary, by considering the analytic continuation of the temperature to the complex plane, the partition function can be written as a polynomial in a variable $z \equiv e^{-\beta\epsilon}$, where $\beta = 1/k_BT$ is the inverse temperature and ϵ is a proper energy interval, whose roots contain all thermodynamic information. Formally, the partition function can be written as

$$Z = \sum_E g(E)e^{-\beta E} = e^{-\beta\epsilon_0}\sum_n g_n\left(e^{-\beta\epsilon}\right)^n = e^{-\beta\epsilon_0}\sum_n g_n z^n, \tag{1}$$

where $g(E)$ is the number of states with energy E and E can be set to $E = \epsilon_0 + n\epsilon$. Once the above polynomial can be factorized in a set of complex roots, z_n, all information present in the partition function is encoded in this set. In particular, the expected non-analytic behavior of the free energy, $F = -k_BT\ln Z$, in the vicinity of a phase transition is expected to be recovered by the presence of a real positive zero, i.e., $Z = 0$ for $T = T_c$ in a phase transition. This zero has small imaginary part in finite systems and touches the real axis in the thermodynamic limit. It is called dominant or leading zero.

The main problems in the Fisher zeros approach are the requirement of the entire density of states, $g(E)$, the high polynomial degree (the number of energy levels), and the elevated values of the coefficients of the polynomial, making numerical methods to find the roots unreliable. Actually, the polynomial degree easily surpasses 50,000 while g(E) may span over more than a hundred thousand orders of magnitude. In this scenario, there is no efficient numerical method to determine the partition function zeros. In order to circumvent this difficulty, in Ref. [4], some of us showed that by rescaling the zeros, the polynomial degree and the range of coefficients can be safely reduced without relevant modifications in the position of the leading zero.

The proposed polynomial[4] can be obtained by multiplying eq. 2 by $1 = e^{-\beta_0 E}e^{\beta_0 E}$. Then,

$$Z_{\beta_0} = \sum_E g(E)e^{-\beta_0 E}e^{-(\beta-\beta_0)E} = e^{-\beta\epsilon_0}\sum_n (H_{\beta_0})_n\left(e^{-\Delta\beta\epsilon}\right)^n = e^{-\beta\epsilon_0}\sum_n (H_{\beta_0})_n x^n, \tag{2}$$

where $x \equiv e^{-(\beta-\beta_0)\epsilon}$ and $H_{\beta_0}(E) \equiv g(E)e^{-\beta_0 E}$ is the non-normalized energy probability distribution function at (inverse) temperature β_0. Since $x = e^{\beta_0\epsilon}z$, up to this point only a rescaling of the Fisher zeros was done. The key point of the method[4] is that now we have a clear criterion to filter the most relevant zeros, specially those that indicate a phase transition. For temperatures near the transition temperature, i.e. for $\beta \approx \beta_c$, states with very low probability to occur are not expected to play any significant role in the transition. Then, one might expect that they do not contribute appreciably to the location of the leading zero, in such a way that even discarding those energy states precise estimates for the leading zero can be obtained.

To illustrate that the polynomial can be safely filtered, we refer to Fig. 1. This figure was obtained using exact results[7] for the density of states for the 2D Ising model in a 16×16 square lattice to build the EPD at $\beta_0 = 0.44$. The black squares are the rescaled Fisher zeros, i.e. the energy probability distribution (EPD) zeros considering all states, obtained using Mathematica®. The polynomial degree in this case is 257 and the coefficients span over 76 orders of magnitude! As can be seen, the solver fails to find all roots. As all coefficients are real positive numbers, there should be no root in the positive real axis, but many of them were obtained. Then, setting the maximum value of the coefficients (histogram values) to 1 and discarding values smaller than a given threshold, $h_c = 10^{-2}$, for example, the polynomial degree and coefficients range are drastically reduced. This result is shown by orange diamonds in Fig. 1. The polynomial degree was reduced to 63 and coefficients span over only 2 orders of magnitude. Interesting is the fact that the zero that approaches the most the real positive axis is at the same position for all cut-offs shown. This is the leading zero.

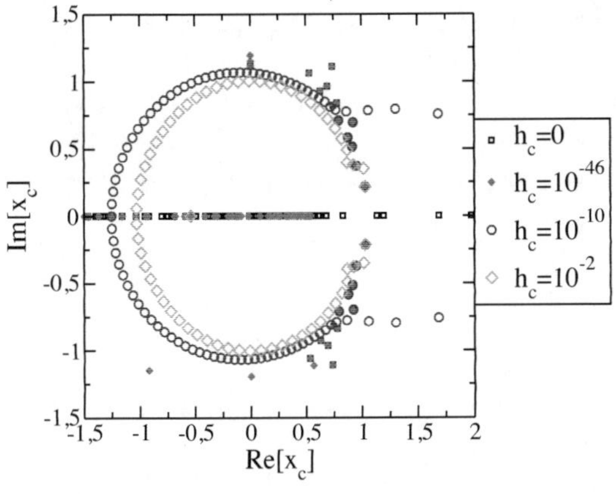

Figure 1. Energy probability distribution (EPD) zeros for the 16×16 2D Ising model at $\beta_0 = 0.44$ considering different cut-offs, h_c, in the histogramas indicated in the legend. As can be seen, the polynomial order is greatly reduced without noticeable modifications to the location of the leading zero.

In addition to the above discussion, note that for $\beta_0 = \beta_c$, where β_c is the critical (inverse) temperature, the leading zero ($x \equiv e^{-(\beta-\beta_0)\epsilon}$) would be at the point $(1, 0)$ in the thermodynamic limit. For finite systems, a small imaginary part is expected. In addition, for β_0 not too close to β_c further deviations are expected since states relevant to the phase transition are not properly sampled. Then, as the critical temperature is approached, initial guesses for β_c can be improved. This allows the application of the following algorithm[4] to locate the transition temperature:

(i) Build a single histogram $H_{\beta_0^j}$ at β_0^j and apply a cut-off if needed.

(ii) Find the zeros of the polynomial with coefficients given by $H_{\beta_0^j}$.

(iii) Find the dominant zero, x_c^j.
  a) If x_c^j is close enough to the point $(1, 0)$, stop.
  b) Else, make $\beta_0^{j+1} = -\frac{\ln(\Re e[x_c^j])}{\epsilon} + \beta_0^j$ and go back to (i).

In step (iii), the dominant zero can be found, for example, by finding the zero nearest to the point $(1, 0)$. As shown in Ref. [4] this algorithm successfully finds the transition temperature for continuous, discontinuous and BKT phase transitions.

3. Application details

The use of the above algorithm to study any system where the EPD can be estimated at a given temperature is straightforward. To illustrate it in more detail, consider the 3-state Potts model[8] in a square lattice, in which at each lattice site, i, a discrete spin variable, $\sigma_i = 1, 2, 3$ is defined. Interactions among neighboring spins are given by:

$$H = -J \sum_{<i,j>} \delta_{\sigma_i,\sigma_j}, \tag{3}$$

where J is a coupling constant and $\delta_{\sigma_i,\sigma_j} = 1$ if $\sigma_i = \sigma_j$ and $\delta_{\sigma_i,\sigma_j} = 0$ otherwise. This model has a second order phase transition at $T_c = \frac{1}{\ln(1+\sqrt{3})} \frac{J}{k_B}$ and critical exponent $\nu = 6/5$ (see Ref. [8]). Good estimates of the EPD can be obtained by conventional Monte Carlo simulations. Here we show results from a simulation using the Metropolis algorithm[1] for a 20x20 lattice with 5×10^4 Monte Carlo Steps (MCS) for thermalization and 5×10^5 MCS to build the EPD. The latter is easily obtained in a simulation by counting how many times each energy state is visited and dividing the result by its largest value in order to obtain a normalized histogram. At this point, one could choose to use all visited energy values or cut off the EPD tails since they are, in general, not very well sampled. Remember that since we deal with a finite number of MCS, not all possible energy values are visited in a simulation, in such a way that the energy range (polynomial degree) is naturally reduced as compared to the Fisher zeros. However, in order to prevent possible deviations that may arise due to the poorly sampled tails, we opted to apply a cut-off, discarding values smaller than 10^{-3}. Care should be taken in this process to ensure that all energy values in the considered energy range, $E \in [E_{min}, E_{max}]$, are properly treated. In fact, it may happen to have an intermediate energy value, $E_{min} < E_a < E_{max}$, whose EPD value is smaller than the chosen threshold, i.e., $H(E_a) < 10^{-3}$. We remark that $H(E_a)$ must be kept as a polynomial coefficient in order to obtain correct results. For more details see Ref. [4] Fig. 2. Very similar results are obtained when no cut-off is used and all energy values between the lowest and highest energies sampled in a Monte Carlo simulation are used.

The animated GIF that can be found in the Supplementary Materials illustrates the entire process that leads to an estimate of the critical temperature. We start at $T_0 = 2J/k_B$ ($\beta_0 = 1/2$), building a normalized histogram and cutting off its tails as described above. Using the histogram values as the polynomial coefficients and numerically solving the polynomial, the leading zero is found to be the zero nearest to the point $(1, 0)$. From the real part of the leading zero, the new estimate of the inverse critical temperature, β_1, is obtained from

$$\beta_1 = -\frac{\ln\left(\Re e[x_c]\right)}{\epsilon} + \beta_0, \tag{4}$$

where $\Re e\{x_c\}$ is the real part of the leading zero and ϵ is the energy step ($E = \epsilon_0 + n\epsilon$). Then, a new simulation is done at $T_1 = 1/\beta_1$, leading to a new histogram from which a new estimate for the critical temperature can be obtained. This process is repeated until the real part of the leading zero is close enough to 1. How close to 1 one should stop depends on the desired accuracy for the critical temperature. The resulting accuracy also depends on the accuracy of the EPD estimate, on the cut-off size and on the zeros finder accuracy. It seems that the EPD quality is the main quantity to focus on in order to improve the critical temperature estimate.

3.1. Convergence issues

Solving a high degree polynomial equation is a complicated task even for the most sophisticated algorithms available, e.g., LAPACK and Mathematica®. As shown in figure 1, problems were found even for a relatively small polynomial (degree 257). In addition, even using the EPD technique to filter the most relevant zeros, it is still possible to encounter very high degree polynomials, since the energy range grows with the system volume. Hence, a loss of accuracy in the dominant zero location may be expected for large systems, which may cause a few problems for the algorithm in its simplest form to find the transition temperature. The remaining of this section is devoted to show very specific cases in which this loss of accuracy was found to cause problems.

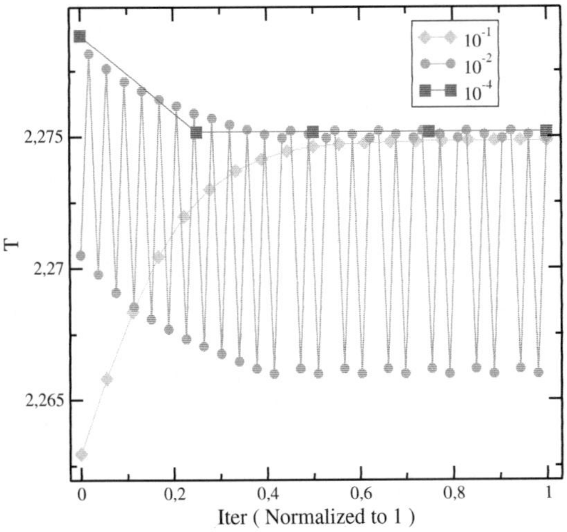

Figure 2. Convergence towards the critical temperature for the 2D Ising model using the values 10^{-1}, 10^{-2} and 10^{-4} as threshold and conventional Monte Carlo simulations. For the 10^{-2} threshold the algorithm did not converge, instead, it starts to oscillate between two branches.

3.1.1. 2D Ising model Here we report results of conventional Monte Carlo simulations for the 2D Ising model in a square lattice. The simulation was carried out using single spin flips, Metropolis algorithm and 10^8 MCS at three different temperatures (2.26, 2.27 and 2.28) in order to use multiple histogram reweighting[9]. For a 150×150 lattice and a 10^{-2} threshold, the algorithm previously proposed does not fully converge, instead, the estimated critical temperature starts to oscillate between two different branches as shown in Fig. 2. Notice that, for the exact same set of histograms, the algorithm converges, as expected, if we choose the threshold 10^{-1} or 10^{-4}, although a small difference between the two critical temperature estimates can be noticed. The observed difference is expected due to the loss of accuracy related to the different cut-off values used.

In Fig. 3, instead of considering the algorithm convergence, we show the real and imaginary parts of the leading zero, i.e., the zero nearest to the point $(1, 0)$, for three different thresholds, 10^{-1}, 10^{-2} and 10^{-4}, and temperatures between 2.26 and 2.28 in steps of 10^{-4}. As can be seen, for the two cases where proper convergence was observed, 10^{-1} and 10^{-4}, the leading zero follows a continuous path, specially near $\Re e\{x\} = 1$. On the other hand, for a threshold of 10^{-2}, it seems that there are forbidden regions, which were observed to cause the convergence problems.

In order to try to understand what is causing this problem, we take a closer look at the zeros map near the point $(1, 0)$. We expect a map of zeros at the transition temperature to be like in Fig. 4 (a), where the leading zero is clearly the closest to the real axis and to the point $(1, 0)$. However, for a 10^{-2} threshold, we have a different structure of zeros as shown on the remaining panels of the same figure. Actually, Fig. 4 (b) is representative of the small cluster of zeros

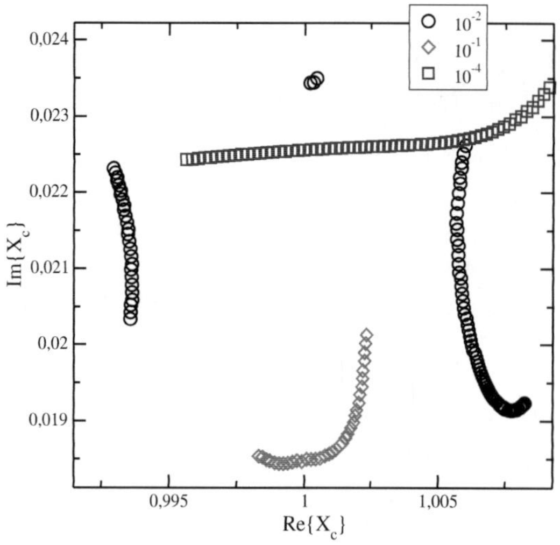

Figure 3. Imaginary and real parts of the EPD leading zero for the 2D Ising model for temperatures $2.26 < T < 2.28$, with $\Delta T = 10^{-4}$ and three different cut-off values.

shown near $\Re e\{x\} = 1$ for the 10^{-2} threshold in Fig. 3, while in (c) and (d) we show typical maps for the left and right branches of Fig.3, respectively. Close inspection of these figures suggest that the leading zero may not be the zero nearest to the point $(1, 0)$ (marked by red circles). Perhaps, the "true" leading zero may be that marked by blue circles in Fig. 4 (c) and (d). We believe that what is causing this misleading choice of zero are little fluctuations on the histogram, specially at its tails. A small variation in a coefficient of a polynomial can lead to huge changes on its zeros location. Other possibility may be a poor accuracy of the zeros finder, which may lead to similar results. Taking into consideration that we have three zeros very close to the point $(1, 0)$, even a small perturbation can affect the choice of the leading zero when using the sole criterion of distance to the point (1,0), preventing proper convergence of the algorithm. We remark, however, that instead of automatically choosing as the leading zero the zero nearest to the point $(1, 0)$, marked with red circles in Fig. 4, one can closely inspect the map of zeros and infer that the leading zero would be the one marked by blue circles. By doing that, proper convergence was recovered in this case.

3.1.2. Artificial spin ice A similar behavior has been observed when applying the method to a completely different system, namely an artificial spin ice model. Artificial spin ices [10] are arrays of single-domain, enlongated magnetic nanoislands designed to mimic the geometric frustration found in pyrochlore spin ice materials [11], which present interesting features such as magnetic monopole-like excitations [12–14]. The interaction between islands is essentially dipolar, and each of them behaves as an effective Ising-like spin, since its magnetic moment is constrained by shape anisotropy to point along the long axis.

We have performed Monte Carlo simulations of a spin ice model with a particular geometry recently proposed and experimentally realized by Wang *et al.* [15], shown in Fig. 5. In our algorithm, the magnetic islands are treated as point-like dipoles and each spin interacts with all other spins in the lattice. The hamiltonian is given by

$$H = \sum_{i>j} \left[\frac{\vec{m}_i \cdot \vec{m}_j - 3(\vec{m}_i \cdot \hat{r}_{ij})(\vec{m}_j \cdot \hat{r}_{ij})}{{r_{ij}}^3} \right], \qquad (5)$$

where $\vec{m}_i$ is the magnetic moment of the ith spin (with $|\vec{m}_i| = \pm 1$), $\hat{r}_{ij}$ is the unit vector that points from spin i to spin j and r_{ij} is the distance between them. The results presented here are for periodic boundary conditions and a lattice size of 32x32 spins.

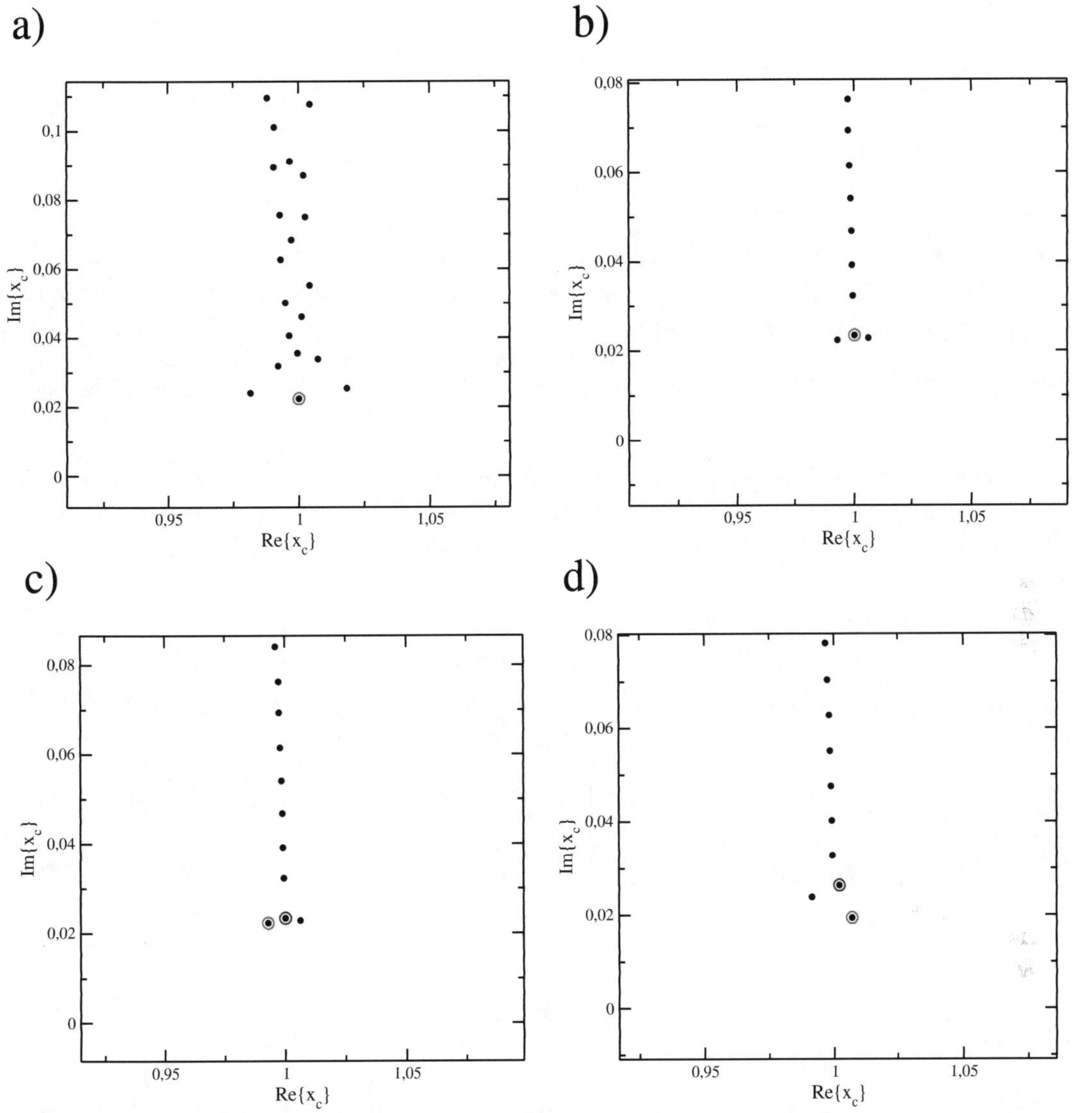

Figure 4. EPD zeros for the 150 × 150 2D Ising model. In (a) we show a typical map for a threshold of 10^{-4}. In (b), (c) and (d) we show maps for a threshold of 10^{-2} at temperatures $T = 2.2752$, $T = 2.2751$ and $T = 2.2662$, respectively. In (b), (c) and (d) we can observe a difference in structure as compared to (a) related to the zeros to the left and to the right of the zero with real part close to 1.

Once the histograms had been obtained, three different cut-off values were used to construct the polynomials and solve for the leading zero. As with the Ising model, one of such cut-off values (10^{-2}) resulted in the algorithm not converging. Instead, as the expected critical temperature is approached, the leading zero drives the algorithm away from it, producing the oscillating pattern observed in Fig. 6. On the other hand, when the cut-off value is set to 10^{-1} or 10^{-3} the algorithm converges, even though the critical temperature indicated is not exactly the same in both cases, as can be expected due to the different accuracy. Fig. 7 illustrates the convergence

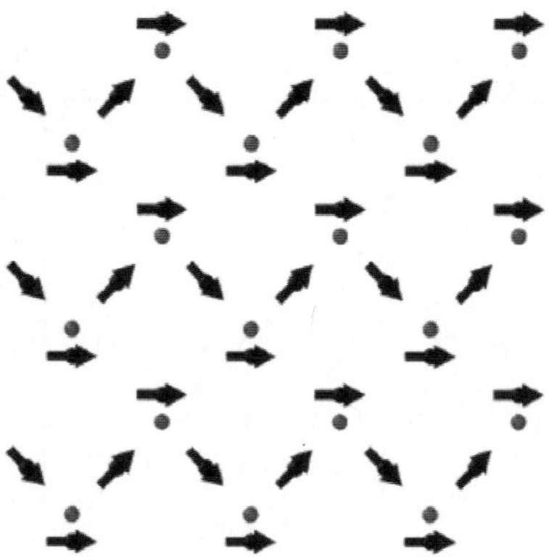

Figure 5. Illustration of the lattice geometry simulated. Green dots are the square lattice sites where collective excitations are expected to live in and the black arrows represent the magnetic islands (treated as Ising spins with dipolar interactions). For more details about this particular geometry see Ref. [15]. Periodic boundary conditions are used. In this image we show a 6x6 lattice.

process. For most cut-off values the real part of the leading zero can get arbitrarily close to one as we approach criticality. Nevertheless, with the 10^{-2} cut-off the leading zeros seem to avoid a certain interval on the real axis, which causes the oscillation shown in Fig. 6.

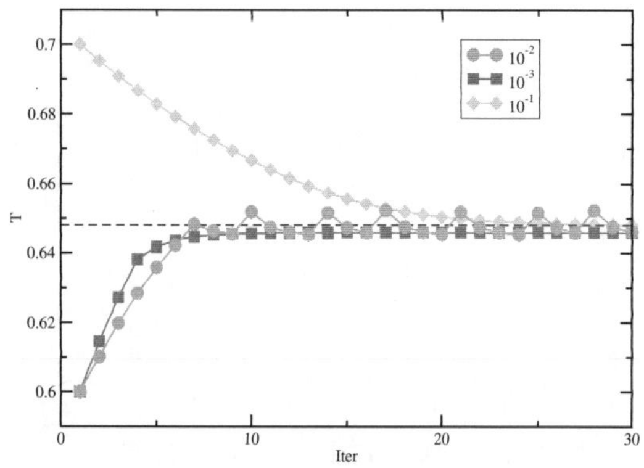

Figure 6. Temperature indicated by the leading zero at each iteration of the algorithm for different histogram cut-off values. For comparison, the horizontal dashed line marks the critical temperature estimated by observing the specific heat peak. The choice of starting temperature does not significantly change the results.

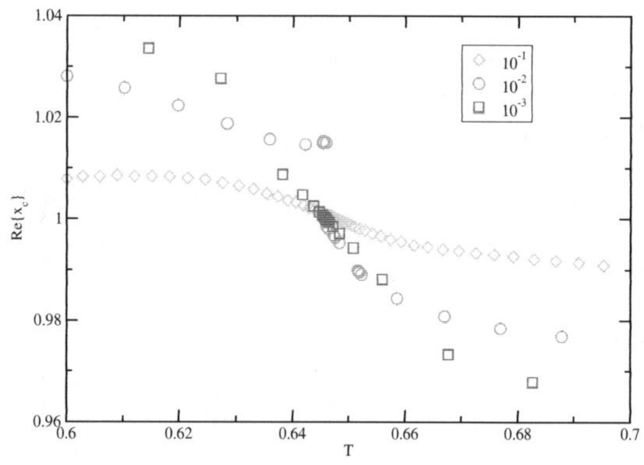

Figure 7. The real part of the leading zero tends to one as the critical temperature is approached, except for the 10^{-2} cut-off, that causes a certain interval on the real axis to be avoided. For each cut-off value, the points shown in the graph are the result of two runs of the algorithm, one starting at a low temperature and the other at a high temperature. We notice that, when the processes converge, the points generated in both runs accumulate near the critical temperature with a real part equal to one. This does not happen with the non-converging processes, in which the leading zeros tend to accumulate elsewhere.

Once again, we systematically searched for the leading zeros at a certain range of temperatures around criticality. The distribution of the zeros on the complex plane is shown in Fig. 8 and is very similar to what had been observed for the Ising model (Fig. 3). For the 10^{-2} cut-off, a gap appears near $Re[X_c] = 1$, while other cut-off values seem to produce zeros that describe a continuous path in this region. In addition, we have found that the maps of zeros are quite similar to that shown in Fig. 4. Once more, we remark that proper convergence of the algorithm was observed when instead of choosing as the leading zero the zero nearest to the point (1, 0) a careful analysis of the zeros map is done.

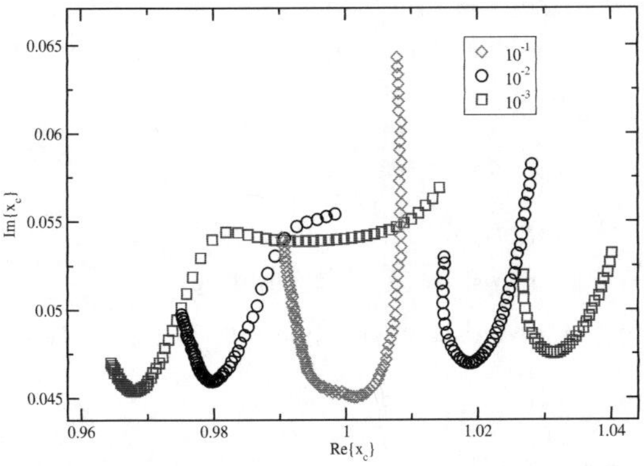

Figure 8. Distribution of the leading zeros in the complex plane for each cut-off value. We have scanned temperatures in the interval $0.6 < T < 0.7$, in steps of 10^{-3}.

4. Closing remarks

In summary, we have shown some details concerning the use of the recently proposed zeros of the energy probability distribution[4] in the study of phase transitions. As shown, by using this method there is no need to define an order parameter to find the transition temperature. In addition, there is no ambiguity in the determination of the pseudo-transition temperature for finite systems and the computational effort to locate the transition temperature is small compared to the use of Fisher zeros and other conventional methods. Indeed, since only partial knowledge of the density of states is required and numerical problems related to the zeros finder's task are greatly reduced without noticeable modifications on the leading zero location, we believe that the energy probability distribution zeros constitute a major improvement in the numerical study of phase transitions. To illustrate its efficiency, we have found for the 3-states Potts model, using the Metropolis algorithm, lattice sizes up to 160 and 10^{-3} for the cut-off of the histograms, $T_c = 0.99491(7)$[4], in excellent agreement with the exact value, 0.99497. For the 2D Ising model, also using Metropolis algorithm, lattice sizes up to 180 and 10^{-2} for the cut-off of the histograms, $T_c = 2.2695(2)$, again in excellent agreement with the exact value, 2.2692. Although some convergence problems were found for the 2D Ising model and an artificial spin ice model, they can be solved by proper inspection of the zeros maps and wise definition of the leading zero. Perhaps, a better definition of which zero is the leading zero or improved accuracy of the zeros finder may solve the observed convergence issues. These questions will be properly addressed in the near future.

Acknowledgments

The authors thank CNPq, CAPES and FAPEMIG, Brazilian agencies, for partial support.

References

[1] Landau D and Binder K 2005 *A Guide to Monte Carlo Simulations in Statistical Physics* (New York, NY, USA: Cambridge University Press)

[2] Privman V 1990 *Finite Size Scaling and Numerical Simulation of Statistical Systems* (World Scientific)

[3] Bachmann M 2014 *Thermodynamics and Statistical Mechanics of Macromolecular Systems* (Cambridge University Press)

[4] Costa B V, Mól L A S and Rocha J C S 2017 *Computer Physics Communications* **216** 77 – 83

[5] Yang C N and Lee T D 1952 *Phys. Rev.* **87**(3) 404–409

[6] Fisher M E 1965 *Lectures in Theoretical Physics: Volume VII C - Statistical Physics, Weak Interactions, Field Theory : Lectures Delivered at the Summer Institute for Theoretical Physics, University of Colorado, Boulder, 1964* v. 7 ed Brittin W (Boulder: University of Colorado Press)

[7] Beale P D 1996 *Phys. Rev. Lett.* **76**(1) 78–81

[8] Wu F Y 1982 *Rev. Mod. Phys.* **54**(1) 235–268

[9] Ferrenberg A M and Swendsen R H 1989 *Phys. Rev. Lett.* **63**(12) 1195–1198

[10] Nisoli C, Moessner R and Schiffer P 2013 *Rev. Mod. Phys.* **85**(4) 1473–1490

[11] Harris M J, Bramwell S T, McMorrow D F, Zeiske T and Godfrey K W 1997 *Phys. Rev. Lett.* **79** 2554–57

[12] Castelnovo C, Moessner R and Sondhi S L 2008 *Nature* **451** 42–5

[13] Mól L A, Silva R L, Silva R C, Pereira A R, Moura-Melo W A and Costa B V 2009 *Journal of Applied Physics* **106** 063913

[14] Mól L A S, Moura-Melo W A and Pereira A R 2010 *Phys. Rev. B* **82**(5) 054434

[15] Wang Y L, Xiao Z L, Snezhko A, Xu J, Ocola L E, Divan R, Pearson J E, Crabtree G W and Kwok W K 2016 *Science* **352** 962–6

IOP Conf. Series: Journal of Physics: Conf. Series **1012** (2018) 012006 doi:10.1088/1742-6596/1012/1/012006

Short–time dynamics of the three–dimensional O(4) model

W G Wanzeller[1] and R S Marques de Carvalho[2]

[1] Universidade Federal da Fronteira Sul - *Campus* Laranjeiras do Sul, BR 158 - Km 406, 85301-970, Laranjeiras do Sul, PR, Brazil

[2] Departamento de Informática em Saúde, Escola Paulista de Medicina, UNIFESP, Rua Botucatu, 862, térreo, 04023-062, São Paulo, SP, Brazil

E-mail: wanderson@uffs.edu.br, marques.carvalho@unifesp.br

Abstract. We study the short-time dynamics of the O(4) spin model in three dimensions. Following the heat-bath time evolution algorithm, we obtain the dynamic critical exponents y and z related to the second moment of the magnetization and to the time correlation length, respectively. We also compute the ratio between the static critical exponents β and ν. Our results are compatible with the ones obtained from other approaches.

1. Introduction

After Janssen et al. [1] and Huse [2] more attention has been drawn to the behavior of the critical dynamics of spin model observables, in particular for the Ising model [3, 4], majority vote [5], damage spreading [6], Potts Model [7] and percolation theory [8] (see [9, 10] for reviews). It is possible to find in the literature numerical measurements of dynamic critical exponents in discrete spin models, but there are no such studies for continuous spin models — except for the three-dimensional Heisenberg model [11] — and for phase transitions in short-time dynamics. In this paper we investigate the short-time phase transition for a continuous spin model. The main idea of this work is to obtain some dynamic critical exponents for the continuous O(4) spin model by using a heat-bath time evolution algorithm. The concept of using the heat-bath method is to update the whole lattice, site by site, changing each spin to a new value without considering the previous spin configuration. The new spin state is selected taking into account its Boltzmann weight. After some iterations all spins in the lattice will obey a Boltzmann distribution and form the heat-bath.

2. Continuous-spin models

O(N) spin models (or nonlinear sigma models) are defined by

$$\mathcal{H} = -J \sum_{\langle i,j \rangle} \vec{S}_i \cdot \vec{S}_j - \vec{H} \cdot \sum_i \vec{S}_i \,, \tag{1}$$

where i and j are the nearest-neighbors sites on a d-dimensional lattice, $\vec{S}_k = (S_k^1, S_k^2, ..., S_k^N)$ is an N-component unit vector at site k and $\vec{H}$ is an external magnetic field. For $N = 1$ we have the Ising model. The case $N = 2$ describes the liquid Helium superfluid transition and $N = 3$ is the classical ferromagnetic Heisenberg model. In this paper we are interested in the $N = 4$ case because it is expected that the phase transition of two-flavor full-QCD would be on the same universality class as the O(4) models [12]. Therefore, the study of short-time phase transition can contribute to the understanding of the full-QCD phase transition.

We consider the short-time dynamics of the three-dimensional O(4) model without external magnetic field ($\vec{H} = 0$). The total magnetization is given by

$$\mathcal{M} \;=\; \sqrt{M_1^2 + M_2^2 + M_3^2 + M_4^2} \quad \text{where } M_\alpha = \frac{1}{V}\sum_i S_i^\alpha \; (\alpha = 1, 2, 3, 4) \tag{2}$$

and the order parameter is the fourth coordinate (M_4). $V(= L^3)$ is the number of lattice sites (volume).

The auto-correlation function is written as

$$A(t) = \frac{1}{V}\left\langle \sum_i \vec{S}_i(0) \cdot \vec{S}_i(t) \right\rangle \tag{3}$$

where $S_i(0)$ is the initial spin configuration.

3. Short-time scaling

Using renormalization-group theory, it can be shown (see Janssen et al. [1]) that the early time evolution of the order parameter (the magnetization fourth coordinate, M_4, from here on represented just by M) already displays universal critical behavior, given by the $k-th$ moment the order parameter

$$M^{(k)}(t, \epsilon, L, m_0) = b^{-k\beta/\nu} \mathcal{M}^{(k)}(t\, b^{-z}, \epsilon\, b^{1/\nu}, L\, b^{-1}, m_0\, b^{x_0}), \tag{4}$$

where m_0 is the initial magnetization, $\epsilon \equiv (T - T_c)/T_c$, $\mathcal{M}$ is a universal function and b is a scale factor, which can be taken equal to $t^{1/z}$. We thus expect, for $T = T_c$, $k = 1$ and small m_0, a power-law behavior at early times

$$M(t)_{\epsilon \to 0} \sim m_0 t^\theta, \tag{5}$$

with $\theta \equiv (x_0 - \beta/\nu)/z$, where β and ν are static critical exponents, x_0 and z are dynamic critical exponents.

When $m_0 = 0$ the auto-correlation function [Eq. (3)] obeys the power law $A(t, t' = 0) \sim t^{-\lambda}$ and defines the dynamic critical exponent $\lambda \equiv (d/z) - \theta$, where $d(= 3)$ is the lattice dimension. The scaling behavior of the second moment of the magnetization can also be described by Eq. (4). Indeed, for $m_0 = 0$, $\epsilon = 0$, and $k = 2$ we have

$$M^{(2)}(t, L) \;=\; t^{-2\beta/z\nu} \mathcal{M}^{(2)}(t^{-1/z} L) \quad \text{which behaves as } M^{(2)}(t) \sim t^y, \tag{6}$$

where the dynamic critical exponent y is defined as $y \equiv (d - 2\beta/\nu)/z$.

The order parameter presents an anomalous time evolution behavior. At the beginning it evolves following an increasing power law [Eq. (5) with $\theta > 0$]. After reaching the maximum value it decreases in two steps: first, following $M \sim t^{-b\nu/z}$ and then, after a long time, the system reaches the equilibrium value. Then, the spin correlation appears and the magnetization goes to $M \sim \exp(-t/\xi_t)$ for a finite system at $T = T_c$, where ξ_t is the time correlation length, given by $\xi_t \approx L^z$.

For a direct measurement of the critical exponent z one can use the method of Ref. [13]

$$F_2(t) \;=\; \frac{M^{(2)}(t)|_{m_0=0}}{[M(t)]^2|_{m_0=1}} \sim \frac{t^{(d-2\beta/\nu)/z}}{t^{-2\beta/z\nu}} \;=\; t^{d/z}. \tag{7}$$

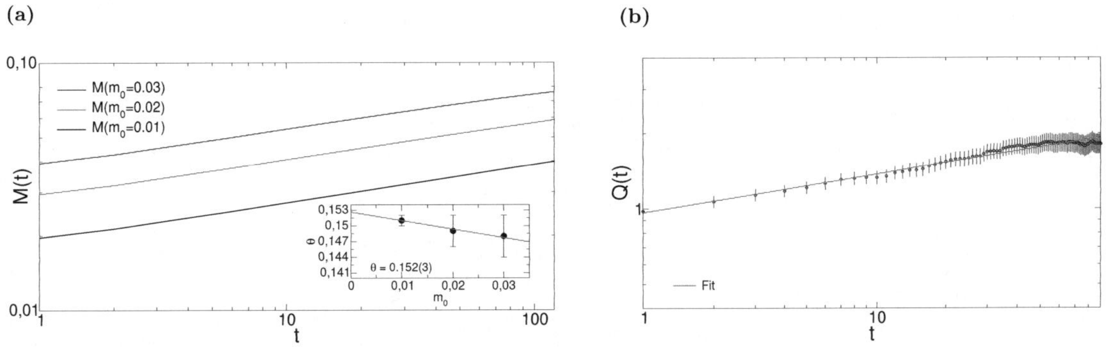

Figure 1: (a) Short-time behavior of the order parameter $M(t)$ [Eq. (5)] for $L = 60$ with different initial conditions ($m_0 = 0.01$, 0.02, and 0.03). Values were averaged over 5000 samples. The inset plot shows the exponent θ as a function of the initial magnetization m_0. The best linear fit gives $\theta = 0.152(3)$ for $m_0 = 0$. (b) Time correlation function [Eq. (8)] for L = 30 and $m_0 = 0$ (not sharp). In this case, we obtained $\theta = 0.153(2)$ with $\chi^2/d.o.f. = 0.16$. The solid line corresponds to the best fit.

In order to avoid the extrapolation and the cumulative error when calculating θ, one can use the time correlation function [14]

$$Q(t) \equiv \frac{1}{V}\left\langle \sum_{i=1}^{V} \sum_{j=1}^{V} \vec{S}_i(t) \cdot \vec{S}_j(0) \right\rangle \sim t^{\theta}, \tag{8}$$

starting from an initial state with zero correlation length and zero (not-sharp) magnetization. This technique was used to study short-time dynamics in the Glauber model, majority vote, "extreme" model [14] and Ising models [15].

The results of our numerical simulations are presented in the following section. For a consistency check, we computed the ratio β/ν from our dynamic critical exponents and compared with the results obtained through other approaches. Some final discussion and conclusions are presented in Section 5.

4. Results

We have studied the three-dimensional O(4) model, performing Monte Carlo simulations with 5,000 seeds (samples or histories) and 100 time steps for each seed. To measure the dynamic critical exponent θ, the simulations were performed at the critical temperature $T_c = 1.0683$ [16] for different values of the initial magnetization ($m_0 = 0.01$, 0.02, and 0.03). Starting from a configuration with null magnetization, we randomly flipped some spins to obtain the desired m_0.

4.1. Dynamic critical exponent θ

The exponent θ was calculated by extrapolating m_0 to zero. For each m_0 we perform Monte Carlo simulations for two lattice sizes, $L = 30$ and 60, corresponding respectively to volumes $V = 30^3$ and 60^3. We have used the heat-bath algorithm [17] with 5,000 seeds and 100 time steps for each seed.

We found a fine power-law behavior of the order parameter in every initial magnetization for both volumes studied, as shown in Fig. 1 (a). The dynamic critical exponent θ, for different values of $m_0 = 0$, is listed in Tab. 1. We have extrapolated $m_0 \to 0$ (last column in the table) for both volumes to obtain $\theta = 0.152(3)$ for the larger size.

Table 1: Dynamic critical exponent θ of the O(4) model. The value for $m_0 = 0$ was obtained by extrapolation.

V	$m_0 = 0.03$	$m_0 = 0.02$	$m_0 = 0.01$	$m_0 = 0.00$
30^3	0.147(1)	0.149(1)	0.150(1)	0.151(6)
60^3	0.148(4)	0.149(3)	0.151(1)	0.152(3)

(a) (b)

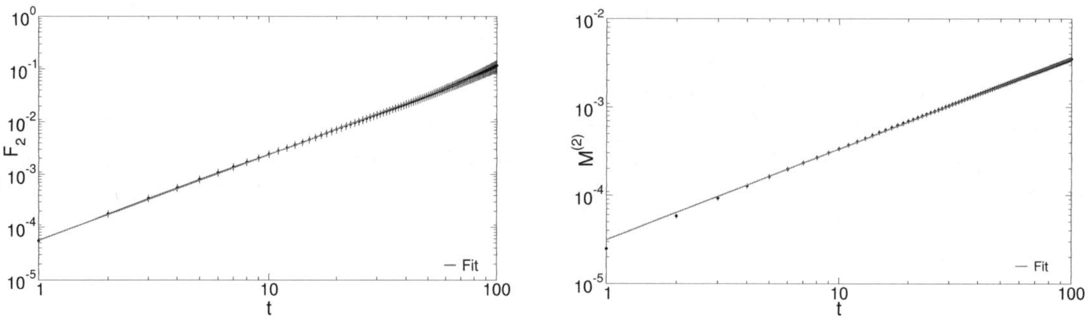

Figure 2: (a) Measurement of dynamic critical exponent $z = 2.176(7)$, $\chi^2/d.o.f. = 0.08$ using the method of [Eq. (7)] with $V = 30^3$ and $5,000$ samples. The best fit in the time interval $[15 : 100]$ is represented by the solid line. (b) Behavior of the second moment of magnetization [Eq. (6)] for $V = 30^3$ and $5,000$ samples. The best fit in the interval $[10 : 90]$ results in $y = 0.908(3)$, $\chi^2/d.o.f. = 0.01$.

The anomalous behavior of the expected order parameter increases with t^θ [Eq. (5)] for very short times and then decreases as $t^{-b\nu/z}$ [Eq. (3)]. On the other hand, although we have a history for each simulation it is not convenient to consider the initial time values in the fit. The power-law of the time evolution just appears after the so called microscopic time (t_{mic}) [4]. In our simulation t_{mic} is around 3 time steps for both adopted volumes. It is the usual value for heat-bath algorithms. The $\chi^2/d.o.f.$ is smaller than 1 for every θ fits. Indeed, in our fits we have not used magnetization values with $t > 72$ when $m_0 = 0.01$, with $t > 51$ when $m_0 = 0.02$ and with $t > 42$ to $m_0 = 0.03$ in both volumes 30^3 and 60^3.

As an alternative to avoid the extrapolation to $m_0 \rightarrow 0$ when measuring θ, one can use the time correlation function [Eq. (8)]. We see in Fig. 1 (b) the behavior of the time correlation function for $V = 30^3$, with 5,000 histories and $m_0 = 0$. The data fit results $\theta = 0.153(2)$ with $\chi^2/d.o.f. = 0.16$.

4.2. Dynamic critical exponents z and y

By using Eq. (7) we calculate the dynamic critical exponent z using $V = 30^3$ and $5,000$ samples. The fit results in $z = 2.176(7)$, with $\chi^2/d.o.f. = 0.08$. In this fit we use the time interval $[25 : 100]$. Both F_2 and the fit results are showed in Fig. 2 (a).

The dynamic critical exponent of the second moment of the magnetization (y) can be calculated from $M^{(2)}(t)|_{m_0=0}$. Indeed, this quantity follows a power law [Eq. (6)] for $m_0 = 0$ and $\epsilon = 0$. Then, using the numerical simulation for F_2, it is possible to measure y. Our data shows $y = 0.908(3)$, with $\chi^2/d.o.f. = 0.01$, in the time interval $[10 : 90]$. The values of $M^{(2)}(t)$ and the best fit can be seen in Fig. 2 (b).

5. Conclusions

We investigated the order parameter short-time behavior of the O(4) continuous spin model. Our results showed a power-law behavior of this observable, with the critical exponent $\theta = 0.152(3)$. For a more precise measurement of the dynamic critical exponent it is advisable to avoid numerical extrapolation. Indeed, it is possible to use the time correlation function. Proceeding this way, we obtained $\theta = 0.153(2)$. Note that, by means of the time correlation function, the final result was improved even using the smaller lattice. Furthermore, the extrapolation (which causes a cumulative error in the final result) was avoided. From the second moment of the magnetization [Eq. (6)] we measured the dynamic critical exponent $y = 0.908(3)$ and, following the method of Ref. [13], we calculated $z = 2.176(7)$.

The short-time power-law behavior of the order parameter, the time correlation function, and the second moment of the magnetization showed evidence of a second order phase transition of the O(4) spin model [9]. The dynamic critical exponents θ, z, and y were obtained with $\chi^2/d.o.f. < 1$. As a consistency check, we have evaluated the ratio β/ν by using the exponents z and y (obtained from our simulations) through the equation $\beta/\nu = (d - zy)/2$. Our estimate is $\beta/\nu = 0.512(5)$. This result agrees with the one obtained through perturbation theory approach, $\beta/\nu = 0.52 \pm 0.02$ [18], and also with the result from Monte Carlo simulations with the single-cluster algorithm technique, $\beta/\nu = 0.5129(11)$ [16].

Acknowledgments

This work was partially supported by CNPq and CAPES. The author (W. G. Wanzeller) thanks T. Mendes (IFSC-USP) for helpful discussions.

References

[1] Janssen H K, Schaub B and Schmittmann B 1989, *Z. Phys.* B **73** 539-549
[2] Huse D A 1989 *Phys. Rev.* B **40** 304-308
[3] Luo H J, Schülke L and Zheng B 1998 *Phys. Rev. Lett.* **81** 180-183; Schülke L and Zheng B 1998 *Phys. Lett.* A **215** 81-85; Li Z B, Schülke L and Zheng B 1995 *Phys. Rev. Lett.* **74** 3396-3398; Zhang J B-, Wang L, Gu D W, Ying H -P and Ji D -R 1999 *Phys. Lett.* A **262** 226-233
[4] Okano K, Schülke L, Yamagishi K and Zheng B 1997 *Nucl. Phys.* B **485** 727-746
[5] Mendes J F F and Santos M A 1998 *Phys. Rev. E* **57** 108-110
[6] Grassberger P 1995 *Physica* A **214** 547-559
[7] Gündüç Y and Aydin M 1996 *Physica A* **232** 265-272; Yaşar F, Gündüç Y, Aydin M and Çelik T 1998 *Physica A* **255** 430-438; Arashiro E, Fernandes H A and Drugowich de Felício J R 2009 *Physica A* **388** 4379-4386
[8] Wanzeller W G, Cucchieri A, Krein G and Mendes T 2004 *Braz. J. Phys.* **34** 247-250; Wanzeller W G, Mendes T and Krein G 2006 *Braz. J. Phys.* **36** 657-659; Wanzeller W G, Mendes T and Krein. G 2006 *Phys. Rev. E* **74** 051123-051128; Wanzeller W G, Krein G and Mendes T 2005 *AIP Conf. Proc.* **739** 608-610
[9] Zheng B 1998 *Int. J. Mod. Phys.* B **12** 1419-1484
[10] Albano E V *et al.* 2011 *Rep. Prog. Phys.* **74** 026501-026530
[11] Fernandes H A, da Silva R and Drugowich de Felício J R 2006 *J. Stat. Mech.* P10002
[12] Pisarski R D and Wilczek F 1984 *Phys. Rev.* D **29** 338-341; Engels J and Mendes T 2000 *Nucl. Phys.* B **572** 289-304; Engels J, Holtmann S, Mendes T and Schulze T 2001 *Phys. Lett.* B **514** 299-308
[13] da Silva R, Alves N A and Drugowich de Felício J R 2002 *Phys. Lett.* A **298** 325-329
[14] Tomé T and de Oliveira M J 1998 *Phys. Rev.* E **58** 4242-4245
[15] Tomé T, Simões C S and Drugowich de Felício J R 2001 *Mod. Phys. Lett.* B **15** 1141-1146
[16] Kanaya K and Kaya S 1995 *Phys. Rev.* D **51** 2404-2410
[17] Creutz M 1980 *Phys. Rev. D* **21** 2308-2315; Kennedy A D and Pendleton B J 1985 *Phys. Lett.* B **165** 393
[18] Baker G, Meiron D and Nickel B 1977 *Compilation of 2-pt. and 4-pt. graphs for continuous spin models*, University of Guelph Report, (unpublished); Rajagopal K and Wilczek F 1993 *Nucl. Phys.* B **399** 395-425

IOP Conf. Series: Journal of Physics: Conf. Series **1012** (2018) 012007 doi:10.1088/1742-6596/1012/1/012007

Bending-Stiffness Dependent Generic Structural Transitions of Helical Polymers

Yifan Dai and Michael Bachmann

Soft Matter Systems Research Group, Center for Simulational Physics, Department of Physics and Astronomy, University of Georgia, Athens, GA 30605, USA

E-mail: dai@smsyslab.org, bachmann@smsyslab.org
Homepage: http://www.smsyslab.org

Abstract. We investigate the impact of competing bending and torsion effects upon the formation of structural phases for helical polymers. For this purpose, parallel-tempering replica-exchange simulations of a coarse-grained polymer model were performed and thermodynamic response quantities analyzed. Extremal energetic and structural fluctuations are used as indicators of significant phase changes. A systematic analysis of the phase structure supports earlier results suggesting the relevance of strong bending stiffness for the stabilization of helical phases. In addition, we find a novel phase at intermediate values of the bending stiffness that provides insight into the process of helix-bundle formation.

1. Introduction

Structural transitions of macromolecules can be considered thermodynamic phase transitions on mesoscopic scales. The conventional approach of analyzing fluctuation quantities such as the heat capacity to identify transition points can be directly employed, but for finite systems, no critical or otherwise catastrophic behavior is expected to occur. Therefore, it is more difficult to obtain a general picture of the phase structure in these systems. Nonetheless, much effort has been dedicated to understanding structural transitions in macromolecular systems of finite size in recent years, including the development of advanced statistical analysis methods [1].

In this paper, we discuss results of continued studies of linear, helical polymers. Helix formation can be induced by a torsion potential [2], which simulates the secondary-structure effects hydrogen bonds cause in real polymers. In addition, non-bonded monomers tend to interact with each other via van der Waals forces, which compete with local ordering achieved by the torsion restraints. As it has been shown, the resulting folded structures, which are supposed to be "functional" in biological systems (such as proteins), do not exhibit any order and look rather like amorphous solids. Apparently, an additional energetic contribution is needed to stabilize low-entropy helical structures such as helix bundles. It was found that a bending potential of two successive bonds indeed helps stabilize the ordered structures but still allows for tertiary structure effects like the formation of compact helix bundles [3, 4, 5].

Here, we investigate structural phase changes for a scenario, in which three-helix structures dominate in the ordered phase, stabilized by strong bending restraints. We systematically weaken the restraints and study the crossover from helical to amorphous compact structures, which govern the conformational behavior at low temperatures if bending restraints are absent.

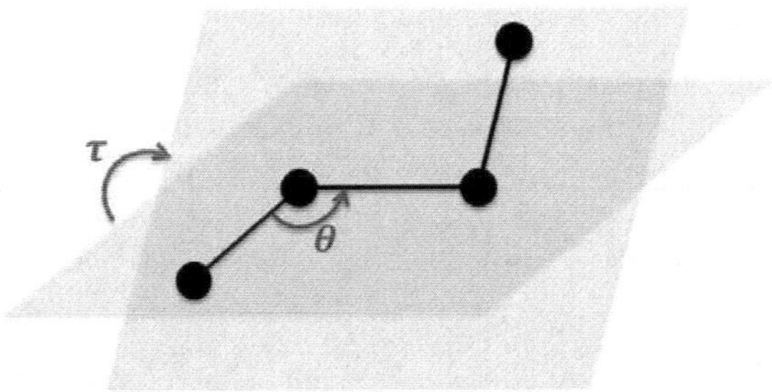

Figure 1. Four monomers are connected by three successive bonds. The bending angle is denoted by θ (blue) and the torsion angle by τ (red).

For this purpose, extensive parallel-tempering Monte Carlo simulations of a coarse-grained helical polymer model were performed [6, 7, 8].

2. Modeling and Simulation

2.1. Modeling

Complex macromolecules such as functional bioproteins are finite systems, which typically consist of thousands of atoms. However, the general structural behavior of macromolecules does not necessarily depend on atomistic details and the chemical group(s) forming a monomer can be considered a coarse-grained entity. Therefore, we employ a generic coarse-grained bead-spring model for an elastic homopolymer, in which individual monomers or chain segments possess only a small number of physical properties. Here, we focus on a model of a polymer chain with $N = 40$ monomers. It already exhibits a distinguished structural phase space, which makes it an ideal candidate for this systematic study.

The conformation of the polymer can be represented by the vector $\mathbf{X} = (\mathbf{x}_1, \mathbf{x}_2, ..., \mathbf{x}_{40})$ where $\mathbf{x}_i$ is the position of the ith monomer. The energy of the polymer chain includes non-bonded, bonded, bending, and torsional interactions. By combining all four potentials, a polymer with conformation $\mathbf{X}$ has the total energy [3, 4]

$$E(\mathbf{X}) = S_{\mathrm{LJ}} \sum_{i>j+1} U_{\mathrm{LJ}}(r_{ij}) + S_{\mathrm{FENE}} \sum_i U_{\mathrm{FENE}}(r_{ii+1}) + S_{\mathrm{bend}} \sum_l U_{\mathrm{bend}}(\theta_l) + S_{\mathrm{tor}} \sum_k U_{\mathrm{tor}}(\tau_k). \quad (1)$$

The interactions between all non-bonded monomers are represented by the dimensionless shifted Lennard-Jones (LJ) potential with cutoff,

$$U_{\mathrm{LJ}}(r) = \begin{cases} 4[(\sigma/r)^{12} - (\sigma/r)^6] - U_{\mathrm{LJ}}(r_c), & r \le r_c, \\ 0, & \text{otherwise}. \end{cases} \quad (2)$$

The van der Waals radius of a monomer is $\sigma = r_0/2^{1/6}$, where r_0 is the inter-monomer distance associated with the minimum of the LJ potential. We use r_0 as the basic length scale and set it to unity in the simulation. The cutoff distance $r_c = 2.5\sigma$ is introduced and the potential is shifted by $U_{\mathrm{LJ}}(r_c) = 4[(\sigma/r_c)^{12} - (\sigma/r_c)^6] \approx -0.0163169$ to avoid a discontinuity at the cutoff point. The non-bonded interaction strength S_{LJ} ($\equiv 1$ in the simulations) provides the reference energy scale for all energetic quantities.

Any pair of bonded monomers interact via the finitely extensible nonlinear elastic (FENE) potential [9, 10]

$$U_{\mathrm{FENE}}(r) = \ln\{1 - [(r - r_0)/R]^2\}, \quad (3)$$

where R is the parameter that limits minimum and maximum bond lengths. The FENE potential diverges as $r \to r_0 \pm R$. The bond interaction strength is $S_{\mathrm{FENE}} = -KR^2/2$ for which we use standard parameter values $K = (98/5)r_0^2$ and $R = (3/7)r_0$ [4].

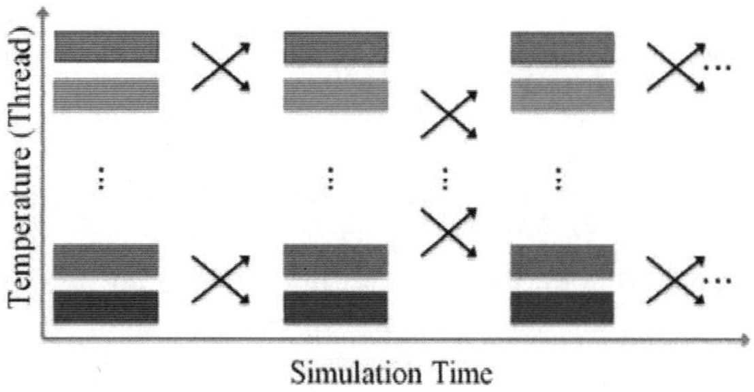

Figure 2. Alternating replica-exchange scheme. Red boxes represent high-temperature and blue boxes low-temperature threads. Arrows indicate the exchanges between neighbors. The first and last thread are idle during every other exchange.

The bending energy is given by

$$U_{\mathrm{bend}}(\theta) = 1 - \cos(\theta - \theta_0), \tag{4}$$

where θ is the bending angle formed by two successive bonds (see Fig. 1). We set the reference bending angle to $\theta_0 = 1.742$.

Similarly, the torsional energy is given by [2]

$$U_{\mathrm{tor}}(\tau) = 1 - \cos(\tau - \tau_0), \tag{5}$$

where τ is an out-of-plane angle formed by three successive bonds as shown in Fig. 1. The reference torsional angle is $\tau = 0.873$. With these preferences of θ_0 and τ_0, the polymer in our model resembles α-helices with about 4 monomers per turn in the ground state.

In recent studies, the structural phases for flexible ($S_{\mathrm{bend}} = 0$) and semiflexible ($S_{\mathrm{bend}} = 200$) polymers could be identified for a large array of torsion strengths S_{tor} [3, 4, 5]. However, the structural transition behavior for bending strengths S_{bend} between these limits has not yet been investigated systematically. We know from previous studies that there are major differences in the structural behavior of flexible and semiflexible polymers at low temperatures. Low-energy conformations of flexible polymers with 40 monomers are compact for $S_{\mathrm{tor}} = 5$ and do not exhibit striking features of order. These structures were classified as amorphous solids. Contrarily, under the same conditions semiflexible polymers form 3-helix bundles, which resemble tertiary structures composed of helical segments as found in proteins. Upon heating, the compact structures melt and globular conformations dominate the "liquid" phase, whereas an additional liquid-gas transition destroys any remaining local structural order. Random-coil conformations govern the "gas" phase.

In this study, we systematically vary S_{bend} at constant torsional strength $S_{\mathrm{tor}} = 5$ and investigate the crossover regime between the structural phases of flexible and semiflexible helical polymers described above.

2.2. Simulation and sampling

The standard importance-sampling technique typically used in Monte Carlo simulations is the Metropolis method, which is based on the canonical microstate probability $p(\mathbf{X}) \sim e^{-\beta E(\mathbf{X})}$ at the given temperature T ($\beta = 1/k_{\mathrm{B}}T$). Thus the acceptance probability $P^{\mathrm{MC}}_{\mathrm{accept}}$ is governed by the ratio of the canonical weights of the microstates

$$P^{\mathrm{MC}}_{\mathrm{accept}}(\mathbf{X} \to \mathbf{X}') = \min\left(1, e^{-\beta\Delta E}\right), \tag{6}$$

where $\Delta E = E(\mathbf{X}') - E(\mathbf{X})$. A Monte Carlo displacement update from $\mathbf{X}$ to $\mathbf{X}'$ is proposed within a cubic box with edge lengths $d = 0.3r_0$. If the energy of the new microstate in each Cartesian direction is smaller after an update, $E(\mathbf{X}') < E(\mathbf{X})$, the move is accepted. If the

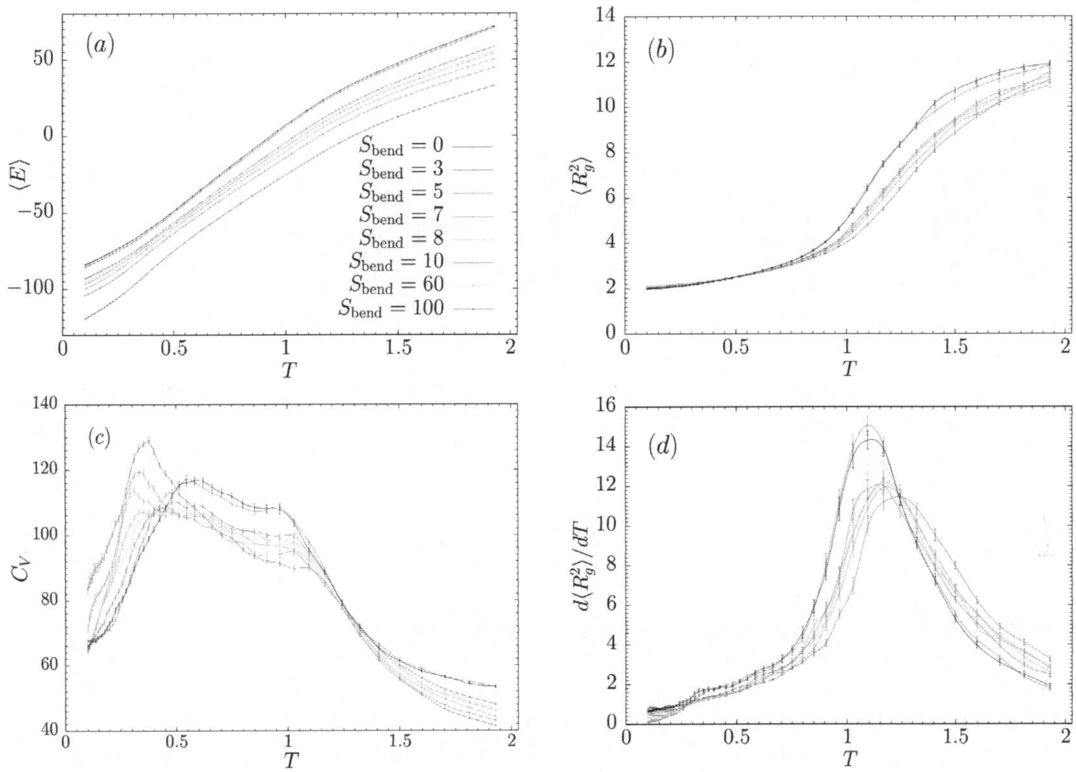

Figure 3. Mean values of (a) energy $\langle E \rangle$, (b) squared gyration radius $\langle R_g^2 \rangle$ and the corresponding fluctuation widths (c) heat capacity C_V, (d) $d\langle R_g^2 \rangle / dT$ as functions of temperature T at various bending strengths S_{bend} for a helical polymer with 40 monomers and torsion strength $S_{\text{tor}} = 5$.

chosen update provokes an increase of energy, $E(\mathbf{X}') > E(\mathbf{X})$, the conformational change is accepted only with the probability $e^{-\beta \Delta E}$. However, the Metropolis method is also one of the least efficient methods due to the difficulty to overcome local free-energy barriers within a reasonable number of local moves. In order to enhance the sampling of low-energy configurations and improve the efficiency, we employ the replica-exchange Monte Carlo (parallel tempering) method [6, 7, 8]. In parallel tempering, multiple process threads are utilized where each thread corresponds to one temperature. After a sufficient number of Monte Carlo sweeps (~ 2000) have been performed, the configuration in each temperature thread is attempted to be exchanged with one of its neighbors. Since the first and last thread have only one neighbor, they remain idle among half of the exchanges. The exchange scheme is shown in Fig. 2. In our study we use 48 temperature threads in the interval $T \in [0.1, 2)$ where $T_i = 1.065\, T_{i-1}$. The acceptance probability $P_{\text{accept}}^{\text{PT}}$ for a single exchange between thread i and thread j is

$$P_{\text{accept}}^{\text{PT}} = \min\left(1, e^{-(\beta_i - \beta_j)[E(\mathbf{X}_j) - E(\mathbf{X}_i)]}\right). \quad (7)$$

Statistical errors of the quantities measured in the individual threads are calculated using the Jackknife error estimation method (for a description see, e.g., Ref. [1]).

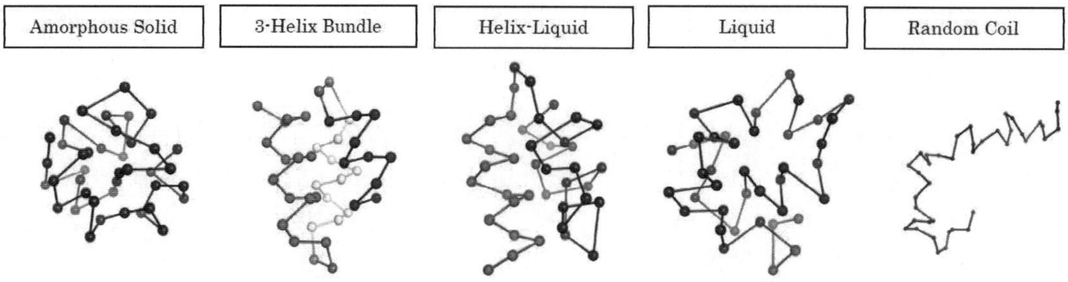

Figure 4. Representative conformations dominating the different structural phases identified for bending strengths in the interval $0 < S_{\text{bend}} < 200$ at $S_{\text{tor}} = 5$.

3. Results

For the investigation of the thermodynamic behavior of helical polymers with different bending strengths, we employ conventional canonical statistical analysis. Energetic and structural thermodynamic quantities were measured at the temperatures used in the parallel-tempering simulation and characteristic features such as inflection points, peaks, and "shoulders" in the curves are used as indicators of structural transitions. Because the system is finite, discontinuities do not occur and the temperatures, at which a specific transition signal is identified in different quantities, do not typically coincide. Therefore, the thermodynamic interpretation of the results is more qualitative than quantitative. Nonetheless, this approach is capable of revealing the most significant features of the transition behavior and the dependence on the bending stiffness.

The results for the general thermodynamic quantities obtained in our simulations at $S_{\text{tor}} = 5$ are compiled in Fig. 3. Shown are the mean energy $\langle E \rangle$ and mean squared radius of gyration $\langle R_g^2 \rangle$, as well as the corresponding thermal fluctuations $C_V = d\langle E \rangle/dT$ (heat capacity) and $d\langle R_g^2 \rangle/dT$ as functions of temperature for different bending strengths $S_{\text{bend}} \in [0, 100]$.

The limit $S_{\text{bend}} = 0$ is associated with flexible polymers and our results are in agreement with previous studies [4]. On the other end, $S_{\text{bend}} = 100$ represents a semiflexible helical polymer with strong bending restraints which stabilize helical conformations. For $S_{\text{tor}} = 5$ and $S_{\text{bend}} = 100$, the helical phase is dominated by 3-helix bundles. No noteworthy change in transition behavior occurs for larger bending strengths $S_{\text{bend}} > 100$. It is, therefore, appealing to take a closer look at intermediate values of S_{bend} and study the change in transition behavior when crossing over from flexible to semiflexible polymers.

The heat capacity shown in Fig. 3(c) exhibits two major transition signals for all bending strengths. The peaks near $T \approx 1$ (with small variations depending on S_{bend}) are supported by strong fluctuations in the radius of gyration [see Fig. 3(d)]. This is an indication that the transition causes major changes in the spatial dimension of the structures and can clearly be associated with the gas-liquid transition between globular and random, expanded conformations. The liquid phase of helical polymers is characterized by globular, overall disordered structures, which may contain short, locally ordered segments such as helical turns. The random-coil conformations do not exhibit any indication of order and the fluctuation behavior of the individual monomers is only constrained by the bond potential.

The other transition signal in the specific-heat curves, which for the different S_{bend} values is found in the interval $0.3 < T < 0.6$, is only weakly reflected in the fluctuations of the radius of gyration. This indicates that it is a transition between compact structures: the globular conformations above the transition temperature and the more ordered, but not necessarily

IOP Conf. Series: Journal of Physics: Conf. Series **1012** (2018) 012007 doi:10.1088/1742-6596/1012/1/012007

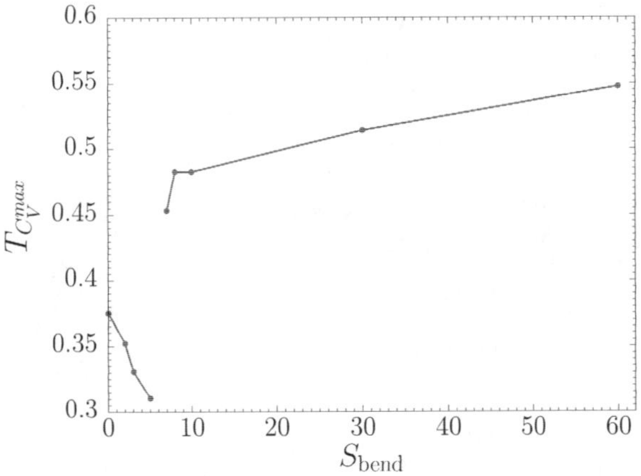

Figure 5. Dependence of the peak temperature of the heat capacity $T_{C_V^{\max}}$ on the bending strength S_{bend} for the compact phases.

denser, conformations at low temperatures. From the studies in the flexible and semiflexible limits, we know that for S_{tor} the low-energy conformations of flexible polymers are amorphous solids, whereas semiflexible helical polymers form very stable, highly ordered 3-helix bundles. Hence, whereas the transition can generically be referred to as the liquid-solid transition in both cases, the structural features in the solid phase are strikingly different. A representative example of an amorphous solid and a 3-helix bundle are shown in Fig. 4.

In order to understand better how these different solid phases are formed, it is helpful to investigate in more detail polymers with intermediate bending strengths. A closer look at the shift of the liquid-solid transition temperature gives a decisive hint. Using again the peak temperature of the heat capacity curves in the low-temperature regime, we observe that by increasing the bending stiffness S_{bend}, the transition point shifts to *lower* temperatures, while the peak height decreases. Then, at about $S_{\text{bend}} = 5$, the signal has shifted to $T \approx 0.31$, but for $S_{\text{bend}} = 7$ another signal appears at about $T \approx 0.45$. This signal becomes stronger for larger S_{bend} values and shifts to *higher* temperatures. Figure 5 shows how the peak locations change with S_{bend}. The "gap" between the signals is clearly visible. This behavior gives rise to the assumption that an intermediate phase exists for $5 < S_{\text{bend}} < 7$, which mediates the crossover between amorphous and helical structures. Indeed, by visual inspection of the dominant structures, we find a new structure type which has not been observed in the limiting cases of flexible and semiflexible polymers. It is a compact and hybrid "helix-liquid" structure, which possesses an ordered helical segment attached to a disordered ("liquid") body. This intermediate structural phase is rather weak and occurs only for bending strengths $S_{\text{bend}} = 5 \ldots 7$ (at $S_{\text{tor}} = 5$). However, it shows how helical bundles form out of the liquid phase, provided the bending stiffness is sufficiently strong to counter the non-bonded interactions that cause the formation of most compact amorphous (or crystalline) structures.

4. Summmary

We have performed parallel-tempering Monte Carlo simulations of a coarse-grained model of helical polymers, in which non-bonded and bonded pair interactions compete with multi-body bending and torsion potentials. For a specific scenario, in which the solid structural phase of semiflexible polymers is governed by 3-helix bundles and by amorphous, solid structures for flexible polymers, we modify the bending stiffness systematically to understand the significant difference in the shapes of the solid structures. Canonical statistical analysis of energetic and structural fluctuation quantities reveals that another structural phase exists in a certain intermediate interval of bending stiffness values, where conformations are composed of a single helix attached to a disordered "liquid" body. Future studies on the characteristic and quantitative features of this intermediate phase are necessary to understand better the general consequences for folding processes of polymer chains under the influence of competing interactions and constraints.

5. Acknowledgments

We thank Dr. Matthew J. Williams for helpful comments. The project was supported by the National Science Foundation (NSF) under Grant No. DMR-1463241.

References

[1] Bachmann M 2014 *Thermodynamics and Statistical Mechanics of Macromolecular Systems* (Cambridge: Cambridge University Press)
[2] Rapaport D C 2002 *Phys. Rev. E* **66** 011906
[3] Williams M J and Bachmann M 2015 *Phys. Rev. Lett.* **115** 048301
[4] Williams M J and Bachmann M 2016 *Phys. Rev. E* **93** 062501
[5] Williams M J and Bachmann M 2016 *Polymers* **8** 245
[6] Swendsen R H and Wang J S 1986 *Phys. Rev. Lett.* **57** 2607
[7] Geyer C J 1991 *Computing Science and Statistics: Proceedings of the 23rd Symposium on the Interface* edited by Keramidas E M (Fairfax Station, VA: Interface Foundation)
[8] Hukushima K and Nemoto K 1996 *J. Phys. Soc. Jpn.* **65** 1604
[9] Kremer K and Grest G S 1990 *J. Chem. Phys.* **92** 5057
[10] Milchev A, Bhattacharya A, and Binder K 2001 *Macromolecules* **34** 1881

IOP Conf. Series: Journal of Physics: Conf. Series **1012** (2018) 012008 doi:10.1088/1742-6596/1012/1/012008

Effects of Stiffness on Low Energy States in a Lattice Protein Model for Crambin

Alfred C.K. Farris[1], **Guangjie Shi**[1], **Thomas Wüst**[2], **and David P. Landau**[1]

[1] Center for Simulational Physics, The University of Georgia, Athens, GA 30602, USA
[2] Scientific IT Services, ETH Zürich, 8092 Zürich, Switzerland

E-mail: `alfred.farris@uga.edu`

Abstract.
Many studies inspired by the HP lattice protein model have helped to confirm the importance of the hydrophobic "driving force" during folding. Unfortunately, the high level of coarse-graining inherent to this model leads to significant limitations; results from proteins studied under the framework of the HP model fail to reproduce many, sometimes significant, details of the folding process, and the obtained ground states are usually highly degenerate. We propose simple modifications to the original HP model, with the goal of reducing degeneracy and gaining insight into how other interaction parameters influence the folding, while retaining the computational simplicity of lattice models. Namely, we introduce a "neutral" monomer (0) to further divide the hydrophobicity scale and an energetic penalty for "bends" in the protein to account for rigidity. Using replica-exchange Wang-Landau (REWL) sampling and suitable Monte Carlo trial moves, we obtain a unique (non-degenerate) ground state for the new lattice mapping of Crambin (a small, 46 amino acid plant protein), and investigate the effects of stiffness on the folding and the low energy structures.

1. Introduction

Protein folding has been studied from many different perspectives for over half a decade, but many unanswered questions remain; we still do not understand exactly how such complex biomolecules fold so efficiently [1]. The statistical mechanical approach to studying protein folding by way of polymer-based modeling has substantially increased our understanding of these systems [2, 3]. Many studies of, or inspired by, the original hydrophobic-polar (HP) model [4, 5] have highlighted the importance of the hydrophobic "driving force", but we still wish to uncover which other interactions contribute most relevantly to the folding process [3]. Polymers, as well as polymer models of proteins, such as the aforementioned HP model, have been extensively investigated by a variety of clever Monte Carlo methods [6–20], including Wang-Landau sampling [21,22], which has also been used to successfully study many other problems of biological interest: protein sequence mutation [23,24], folding in a confined environment [25,26], and surface adsorption [27–30].

In addition to the hydrophobic effect, previous studies have shown that rigidity plays an important role in the folding process, as well as in the functional responsibilities of biomolecules [31–35]. We have used replica-exchange Wang-Landau sampling and slight modifications to the original HP model to investigate the effect of protein stiffness on low energy states of a lattice protein model for Crambin.

2. Models and Simulation Methods

2.1. Semi-flexible H0P Lattice Protein Model for Crambin

The classic HP model greatly simplifies the complexity of protein folding by classifying each of the twenty amino acids as either hydrophobic (H) or polar (P), based on their tendency to interact with water [4, 5]. The protein is then placed on a lattice, with the H and P monomers occupying the lattice sites, connected via rigid bonds. The polymer then performs a self-avoiding random walk, effectively incorporating excluded volume effects inherent to real proteins. To account for the effective hydrophobic driving force, the HP model assigns each non-bonded, nearest neighbor HH contact an energy, ϵ_{HH}, that lowers the total energy of the system; the configuration(s) with the lowest energy is(are) the ground state(s). One problem inherent to this model is that the ground state energies are degenerate, which stands in contrast to the unique native state found in real proteins. Additionally, this model doesn't account for polymer stiffness, an effect that should not be overlooked [31–34]. Here we propose a few simple modifications to the original HP model, with the goal of reducing degeneracy and gaining insight into how other interaction parameters influence the folding, while retaining the computational simplicity of lattice models. We first introduce a "neutral" monomer (0) to further divide the hydrophobicity scale [36, 37], and an energetic penalty for "bends" in the protein to account for rigidity [35]; we call this the Semi-flexible H0P model, and a sample configuration, along with interaction parameters is seen in Figure 1. The Hamiltonian for our model is given as follows:

$$\mathcal{H} = -\epsilon_{HH} n_{HH} - \epsilon_{H0} n_{H0} - \epsilon_{\theta} n_{\theta}, \tag{1}$$

where n_{HH} is the number of HH contacts, ϵ_{HH} is the energy of one such contact, and likewise for the other parameters. In this study, we set $\epsilon_{HH} = 1$ and $\epsilon_{H0} = 0.5$.

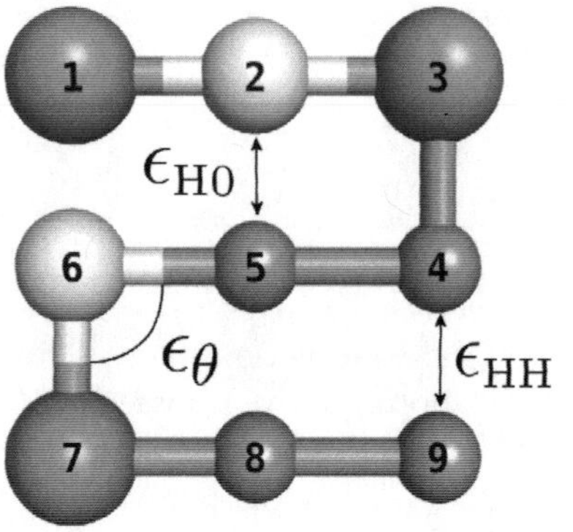

Figure 1. Sample semi-flexible H0P model configuration. Hydrophobic, neutral, and polar monomers are colored in grey, white, and orange, respectively. The interaction between monomers 2 and 5 contributes energy ϵ_{H0}, the interaction between monomers 4 and 9 contributes energy ϵ_{HH}, and the angle formed by monomers 5, 6, and 7 contributes energy ϵ_{θ}. In total, there are 2 HH contacts (n_{HH}), 1 H0 contact (n_{H0}), and 4 bond-angles (n_{θ}).

In this study we investigate Crambin, a slightly hydrophobic protein found in cabbage [38,39]. Due to its relatively small size (46 amino acids), Crambin is one of the more computationally tractable proteins to model. It was previously mapped onto the HP [40] and H0P [37] models with sequence names HP3D46 and H0P3D46, respectively.

2.2. Replica-exchange Wang-Landau Sampling

Traditional Monte Carlo methods, e.g. Metropolis sampling, sample prohibitively slowly at low temperatures and fail for systems with complex free energy landscapes. Wang-Landau sampling [41, 42] is an iterative Monte Carlo method whose ultimate goal is obtaining a system's density of states (DOS), $g(E)$, by performing a random walk in energy space. During the simulation, a trial move to change the system from state A to B is proposed, and accepted with probability

$$P(A \to B) = min\left\{1, \frac{g(E_A)}{g(E_B)}\right\}, \tag{2}$$

where $g(E_A)$ and $g(E_B)$ are the current estimates of the DOS for states A and B, respectively. After a trial move where state n is accepted, $g(E_n)$ is updated via $g(E_n) \to f \cdot g(E_n)$, where f is a modification factor. A histogram $H(E)$ of visited system states is also kept, and updated via $H(E_n) \to H(E_n) + 1$ after state n is accepted. The histogram is checked at predefined intervals, and reset to zero when all system states have been visited roughly equally: in this study, we enforce an "80% flatness criterion", meaning all histogram entries must be at least 80% of the average histogram value. The modification factor is then updated via $f \to \sqrt{f}$, and this process is repeated until $f < f_{min}$, some predefined minimum value.

Replica-Exchange Wang-Landau sampling (REWL) is an extension that combines the power of traditional Wang-Landau sampling with replica-exchange Monte Carlo [43,44]. In REWL, the energy range of a system is split into multiple, overlapping windows, each of which is sampled by one or more independent walkers. A replica-exchange between walkers in neighboring windows i and j is proposed periodically, and accepted with probability

$$P_{acc} = min\left\{1, \frac{g_i(E_i)}{g_i(E_j)} \frac{g_j(E_j)}{g_j(E_i)}\right\}, \tag{3}$$

where $g_i(E_j)$ is walker $i's$ current estimate for the density of states of the configuration proposed by walker j, and likewise for the other terms. This framework allows each replica to travel between windows with different energy ranges, which further helps to reduce trapping in free energy minima. REWL produces significant speed-up in simulations of complex systems such as lattice protein adsorbtion and the Potts model [43,44].

In this study, f was initially set to e^1, and $f_{min} = exp(10^{-8})$. Our trial moves were chosen randomly using the Mersenne Twister random number generator, such that the percentage of each move type were as follows: 75% pull moves [45], 23% bond-rebridging moves [46], and 2% pivot moves [47]. This combination of Monte Carlo moves has proven very efficient in studying lattice proteins together with Wang-Landau sampling [21,22].

2.3. Thermal Properties and Protein Structures

With $g(E)$ obtained from REWL sampling, the partition function (Z) can be computed at any temperature (T) simply as

$$Z(T) = \sum_E g(E)e^{-E/k_BT}, \tag{4}$$

where E is the system energy, and k_B is the Boltzmann constant. We then calculate the statistical mechanical averages of any quantity Q as follows:

$$\langle Q \rangle = \frac{\sum_E Q \cdot g(E)e^{-E/k_BT}}{Z}. \tag{5}$$

The heat capacity (C_V) can be calculated from fluctuations in the energy via

$$C_V(T) = \frac{\langle E^2 \rangle - \langle E \rangle^2}{k_BT^2}. \tag{6}$$

Due to the high degeneracy of most energy levels, as well as the many symmetries associated with the simple cubic lattice, one must be very careful when comparing "different" structures with the same energy. A highly efficient method of searching for, and keeping track of different structures was previously introduced by Shi et. al. [24]. The idea behind this method is to sample configuration space for structures corresponding to a predefined target energy using (replica-exchange) multicanonical sampling (REMUCA) [48, 49], with $g(E)$ from the preceding REWL run as the sampling weight, and translate each structure with this target energy into a sequence of directions (SoD) that accounts for lattice symmetries. During the sampling, each unique SoD is stored in a tree-like data structure, to which newly encountered sequences can be quickly compared. If a given structure is found to be different from all others at that energy, it is added to the tree. This process stops once a certain predefined number of states are found, or if no new structures are found in a certain number of Monte Carlo sweeps. The minimum number of REMUCA sweeps used to calculate structural quantities or obtain ground states for a simulation in this study was 10^9. Statistical error bars for all quantities in this study were obtained from a minimum of five independent runs.

3. Results

3.1. Specific Heat

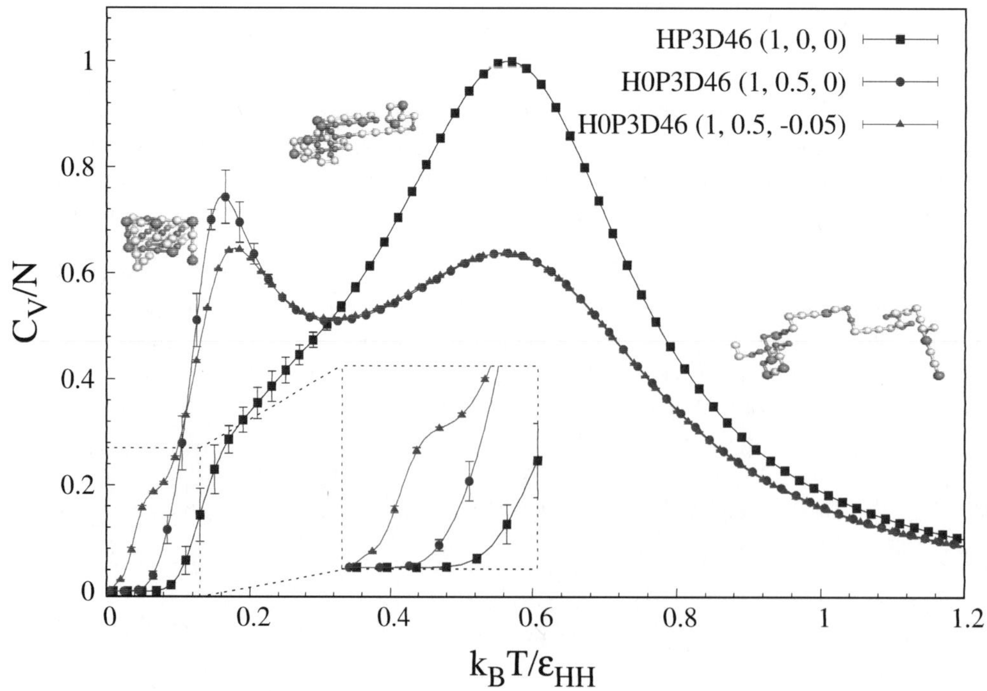

Figure 2. Specific heat curves for the HP, H0P, and semi-flexible H0P model for Crambin, with characteristic structures relative to peak locations. Values in the legend correspond to (ϵ_{HH}, ϵ_{H0}, ϵ_{θ}). Error bars smaller than the points are not shown.

The specific heat offers insight into a system's response to thermal excitations, with peaks indicating substantial "transitions". In smaller proteins, the folding process is thought to occur in two steps [50]: Firstly a coil-globule transition, in which the protein folds from a random coil to a globule. Secondly, the protein's structure becomes even more compact, ultimately resulting in a relaxation to the native state. Figure 2 shows the specific heat curve for the HP, H0P, and semi-flexible H0P ($\epsilon_\theta = -0.05$) variants of Crambin, along with characteristic H0P structures relative to the peak locations.

While the original HP model only begins to hint at a second transition (in the form of a broad shoulder), the H0P and semi-flexible H0P models clearly show at least two transitions, consistent with all-atom simulations of Crambin [51]. More interestingly, however, is the emergence of a third structural transition signal, in the form of a small shoulder at very low temperature when we account for bond-angle stiffness. While this shoulder is quite small, it is clearly larger than statistical error bars.

3.2. Ground State Degeneracies and Low Energy Structures

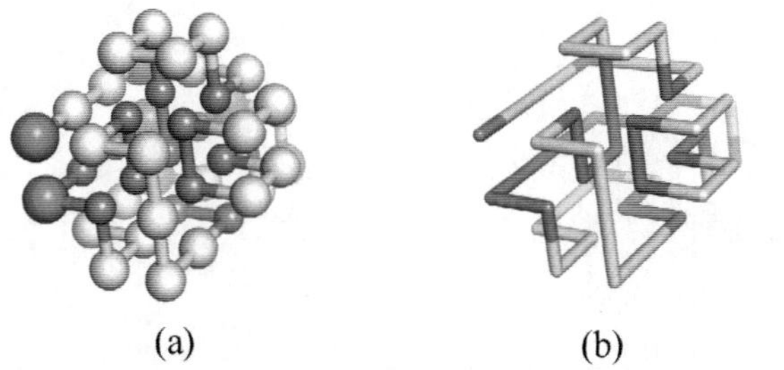

Figure 3. Ground state structure (a) and bond-only representation (b) for the semi-flexible H0P variant of Crambin with $\epsilon_{HH} = 1$, $\epsilon_{H0} = 0.5$, and $\epsilon_\theta = -0.05$.

By introducing the "neutral" monomer type, the H0P model reduces the ground state degeneracy to 121 states from $\sim$ 5.9 x 10^5 states in the HP mapping [37]. Adding a small, but opposing energetic contribution ($\epsilon_\theta = -0.05$) for bond-angles further reduces to only a **single (unique) native state**, consistent with theoretical expectations. The ground state structure is shown in Figure 3. The shoulder that appears at low temperature for this case prompts an investigation into the structures near the ground state.

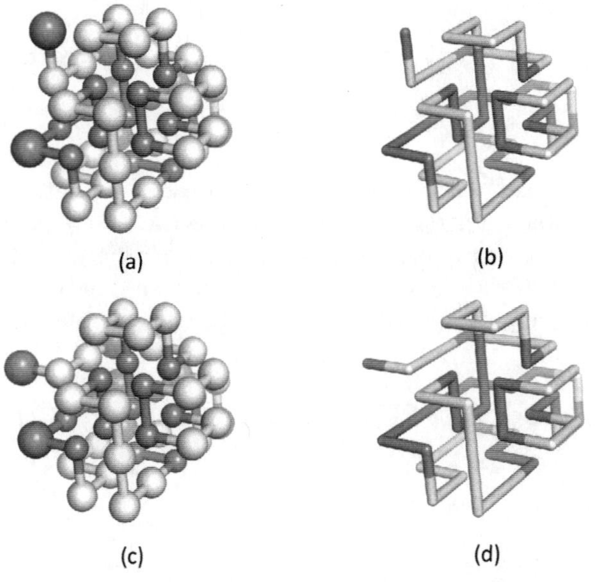

Figure 4. First excited states (a, c) and bond-only representation (b, d) for the semi-flexible H0P variant of Crambin with $\epsilon_{HH} = 1$, $\epsilon_{H0} = 0.5$, and $\epsilon_\theta = -0.05$.

Figure 4 shows the first excited states (a,c), as well as corresponding bond-only representations (b,d). From this figure, it is apparent that the first-excited states are *very* similar to the ground state, with the only difference being an additional "bend" in the tail end of the protein.

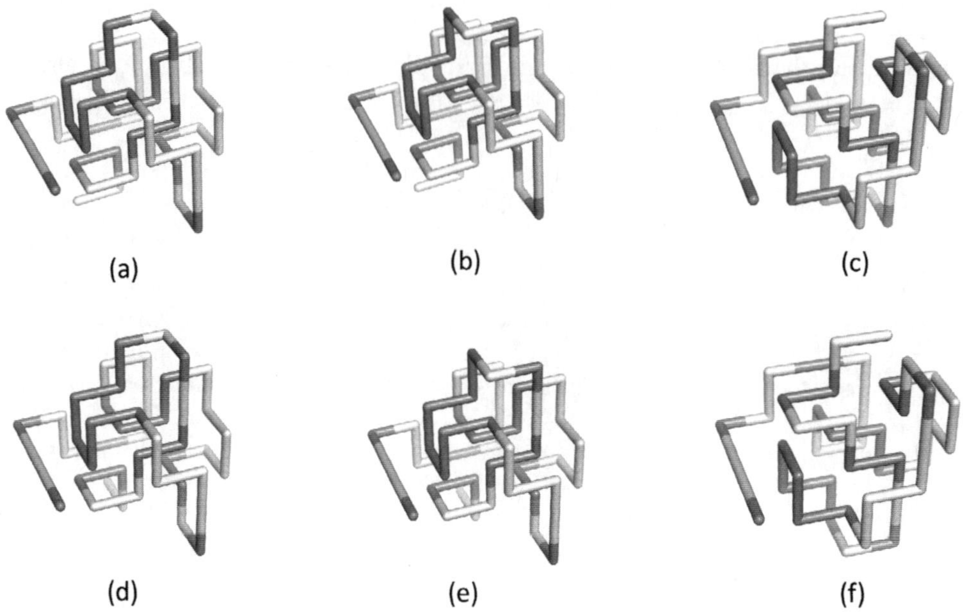

Figure 5. Bond-only representation of the Second excited states for the semi-flexible H0P variant of Crambin with $\epsilon_{HH} = 1$, $\epsilon_{H0} = 0.5$, and $\epsilon_{\theta} = -0.05$.

Figure 5 shows bond-only representations of the six second-excited states. While these structures are very similar to each other, the internal arrangement of amino acids differs *drastically* from the lower energy states. A more quantitative study is underway, but we believe that this internal rearrangement corresponds exactly to the low temperature shoulder observed in the specific heat curve.

4. Conclusion

We have used REWL sampling, along with slight modifications to the original HP lattice protein model to investigate the effects of stiffness on folding. By studying the case where $\epsilon_{\theta} = -0.05$, we reduce the ground state degeneracy by five orders of magnitude from the original HP model, and obtain a non-degenerate ground state. Moreover, we uncover a previously unobserved structural transition at very low temperature. Further investigation of the low energy protein structures reveals a *huge* change in the internal structure between the second and third excited states. A more detailed study is in preparation, in which we aim to uncover the effects of stiffness on the structural properties during folding, as well as to quantitatively describe the internal rearrangement in structure observed between the second and third excited states in our semi-flexible model.

5. Acknowledgements

This study was supported in part by resources and technical expertise from the Georgia Advanced Computing Resource Center, a partnership between the University of Georgia's Office of the Vice President for Research and Office of the Vice President for Information Technology. We would also like to thank the Texas Advanced Computing Resource Center under the NSF XSEDE program, grant no. PHY130014, for computing resources.

References

[1] Dill K A and MacCallum J L 2012 *Science* **338** 1042–6
[2] Kolinski A and Skolnick J 2004 *Polymer* **45** 511–524
[3] Dill K A 1999 *Protein Sci.* **8** 1166–80
[4] Dill K A 1985 *Biochemistry* **24** 1501–9
[5] Lau K F and Dill K A 1989 *Macromolecules* **22** 3986–3997
[6] O'Toole E M and Panagiotopoulos A Z 1992 *J. Chem. Phys.* **97** 8644
[7] Beutler T C and Dill K A 1996 *Protein Sci.* **5** 2037–2043
[8] Grassberger P 1997 *Phys. Rev. E* **56** 3682–3693
[9] Frauenkron H, Bastolla U, Gerstner E, Grassberger P and Nadler W 1998 *Phys. Rev. Lett.* **80** 3149–3152
[10] Bachmann M and Janke W 2003 *Phys. Rev. Lett.* **91** 208105
[11] Hsu H P, Mehra V, Nadler W and Grassberger P 2003 *J. Chem. Phys.* **118** 444
[12] Prellberg T and Krawczyk J 2004 *Phys. Rev. Lett.* **92** 120602
[13] Zhang J, Kou S C and Liu J S 2007 *J. Chem. Phys.* **126** 225101
[14] Liang F and Wong W H 2001 *J. Chem. Phys.* **115** 3374
[15] Zhang J L and Liu J S 2002 *J. Chem. Phys.* **117** 3492
[16] Tang W and Zhou Q 2012 *Phys. Rev. E* **86** 031909
[17] Unger R and Moult J 1993 *J. Mol. Biol.* **231** 75
[18] König R and Dandekar T 1999 *BioSystems* **50** 17–25
[19] Shmygelska A and Hoos H H 2005 *BMC Bioinf.* **6** 30
[20] Yue K, Fiebig K M, Thomas P D, Chan H S, Shakhnovich E I and Dill K A 1995 *PNAS* **92** 325–329
[21] Wüst T and Landau D P 2009 *Phys. Rev. Lett.* **102**(17) 178101
[22] Wüst T and Landau D P 2012 *J. Chem. Phys.* **137** 064903
[23] Holzgräfe C, Irbäck A and Troein C 2011 *J. Chem. Phys.* **135** 195101
[24] Shi G, Vogel T, Wüst T, Li Y W and Landau D P 2014 *Phys. Rev. E* **90** 033307
[25] Ping G, Yuan J M, Vallieres M, Dong H, Sun Z, Wei Y, Li F Y and Lin S H 2003 *J. Chem. Phys.* **118** 8042
[26] Pattanasiri B, Li Y W, Landau D P, Wüst T and Triampo W 2012 *J. Phys.: Conf. Ser.* **402** 012048
[27] Castells V, Yang S and Van Tassel P 2002 *Phys. Rev. E* **65** 031912
[28] Swetnam A and Allen M P 2012 *Phys. Rev. E* **85** 062901
[29] Radhakrishna M, Sharma S and Kumar S K 2012 *J. Chem. Phys.* **136** 114114
[30] Li Y W, Wüst T and Landau D P 2013 *Phys. Rev. E* **87** 012706
[31] Thomas P D and Dill K A 1993 *Protein Science* **2** 2050–2065
[32] Bastolla U and Grassberger P 1997 *J. of Stat. Phys* **89** 1061
[33] Krawczyk J, Owczarek A and Prellberg T 2009 *Physica A* **388** 104–112
[34] Williams M J and Bachmann M 2015 *Phys. Rev. Lett.* **115**(4) 048301
[35] Shi G, Farris A C K, Wüst T and Landau D P 2016 *J. Phys.: Conf. Ser.* **686** 012001
[36] Kyte J and Doolittle R F 1982 *J. Mol. Biol.* **157** 105 – 132
[37] Shi G, Wüst T, Li Y W and Landau D P 2015 *J. Phys.: Conf. Ser.* **640** 012017
[38] Vanetten C, Nielsen H and Peters J 1965 *Phytochemistry* **4** 467 – 473 ISSN 0031-9422
[39] Hendrickson W A and Teeter M M 1981 *Nature* **290** 107 – 113 ISSN 0031-9422
[40] Lattman E E, Fiebig K M and Dill K A 1994 *Biochemistry* **33** 6158–6166
[41] Wang F and Landau D P 2001 *Phys. Rev. Lett.* **86** 2050–2053
[42] Wang F and Landau D P 2001 *Phys. Rev. E* **64** 1–16
[43] Vogel T, Li Y W, Wüst T and Landau D P 2013 *Phys. Rev. Lett.* **110** 210603
[44] Vogel T, Li Y W, Wüst T and Landau D P 2014 *Phys. Rev. E* **90** 023302
[45] Lesh N, Mitzenmacher M and Whitesides S 2003 *RECOMB* 188–195
[46] Deutsch J M 1997 *J. Chem. Phys.* **106** 8849 ISSN 00219606
[47] Sokal A D 1995 *Monte Carlo and Molecular Dynamics Simulations in Polymer Science* (Oxford University Press) edited by K. Binder
[48] Berg B A and Neuhaus T 1991 *Physics Letters B* **267** 249 – 253 ISSN 0370-2693
[49] Berg B A and Neuhaus T 1992 *Phys. Rev. Lett.* **68**(1) 9–12
[50] Jackson S E 1998 *Folding and Design* **3** R81 – R91 ISSN 1359-0278
[51] Shimada J, Kussell E L and Shakhnovich E I 2001 *J. of Mol. Biol.* **308** 79 – 95

IOP Conf. Series: Journal of Physics: Conf. Series **1012** (2018) 012009 doi:10.1088/1742-6596/1012/1/012009

Evolution of tag-based cooperation with emotion on complex networks

F.W.S. Lima

DietrichStauffer Computational Physics Lab, Departamento de Física
Universidade Federal do Piauí, 64049-550,Teresina-PI, Brazil

E-mail: fwslima@gmail.com

Abstract. We study the evolution of the four strategies: Ethnocentric, altruistic, egoistic and cosmopolitan in one community of individuals through Monte Carlo simulations. Interactions and reproduction among computational agents are simulated on undirected Barabási-Albert (UBA) networks and Erdös-Rènyi random graphs (ER).We study the Hammond-Axelrod model on both UBA networks and ER random graphs for the asexual reproduction case. We use a modified version of the traditional Hammond-Axelrod model and we also allow the agents' decisions about one of the strategies to take into account the emotion among their equals. Our simulations showed that egoism and altruism win, differently from other results found in the literature where ethnocentric strategy is common.

1. Introduction

The tag-based cooperation [1, 2, 3, 4, 5, 6, 7, 8] has attracted considerable attention in the scientific community in recent years [9], also among physicists [10, 11, 12, 13, 14].

The scale-free Barabási-Albert (BA) networks[15] and game theory like the Prisoner's Dilemma [16, 17, 18, 19] are important examples of social phenomena simulated by physicists [20]. The scale-free networks have heterogeneity of links among individual agents [15, 21]: A few have many neighbors and most have only few neighbors. The local interactions between computational agents in the prisoner-dilemma game lead to cooperation among similar agents [22], as well as non-cooperation with dissimilar agents (ethnocentrism) [5], even when there is no reciprocity and cooperation is expensive [23]. Sometimes good empirical data exist for comparison with simulations, in particular for political elections [24, 20]. Also economics has been simulated with evolutionary agent models, e.g. [25]. Hammond-Axelrod (HA) models of evolution of ethnocentrism [5] and contingent altruism [22] show that in-group favoritism among similar agents can evolve under a wide variety of conditions, even when there is no bias towards apparently similar agents. Four different types of agents (each labeled with a different color or other tag) compete via prisoner-dilemma-type interactions. For example, ethnocentric agents treat other agents with the same tag better than those with different tags. In addition, a mechanism for inheritance (genetic or cultural) of strategies is included. The ability to discriminate between the in-group and the out-groups through tags and local interactions can overcome egoism and promote cooperative behavior [5]. However, these models [22, 5] use simple square lattices, neglecting thus a plethora of biological or social real-world networks of contacts, complex and mostly heterogeneous [15, 26, 27, 18, 29].

IOP Conf. Series: Journal of Physics: Conf. Series **1012** (2018) 012009 doi:10.1088/1742-6596/1012/1/012009

Lima [28] has studied the HA model of ethnocentrism on both semi-directed and undirected scale-free networks [29, 30, 31]. Lima [28] was interested in how ethnocentrism and egoism emerge spontaneously at the global group-level from the local interactions of differently labeled individual agents. Its itroduced a small change where only agents with reproductive potential α and a costly reduction of its α by c obey the relation $\alpha > c$ can help, thus avoiding negative α. Different from the traditional HA models of evolution of ethnocentrism, their results showed that egoism can win the ethnocentrism for some m agent neighbors ($m = 1$, 2, 5, 10, 20, 50 and 100) [28].

Our objective here is therefore to study the modified HA model of ethnocentrism [28] on both ER random graphs and UBA networks [29, 30, 31] and including now the emotion factor among the agents.

The "emotion" introduced in this paper is the fraction of neighbors of a central node having the same "strategy" as the central node, where strategy is ethnocentric, altruist, egoist or cosmopolitan. For example, the emotion is zero if an altruist finds no other altruists among its neighbors, and is it unity if all its neighbors are altruists. The amount of help given and the amount of cost produced by this help is multiplied by this emotion factor.

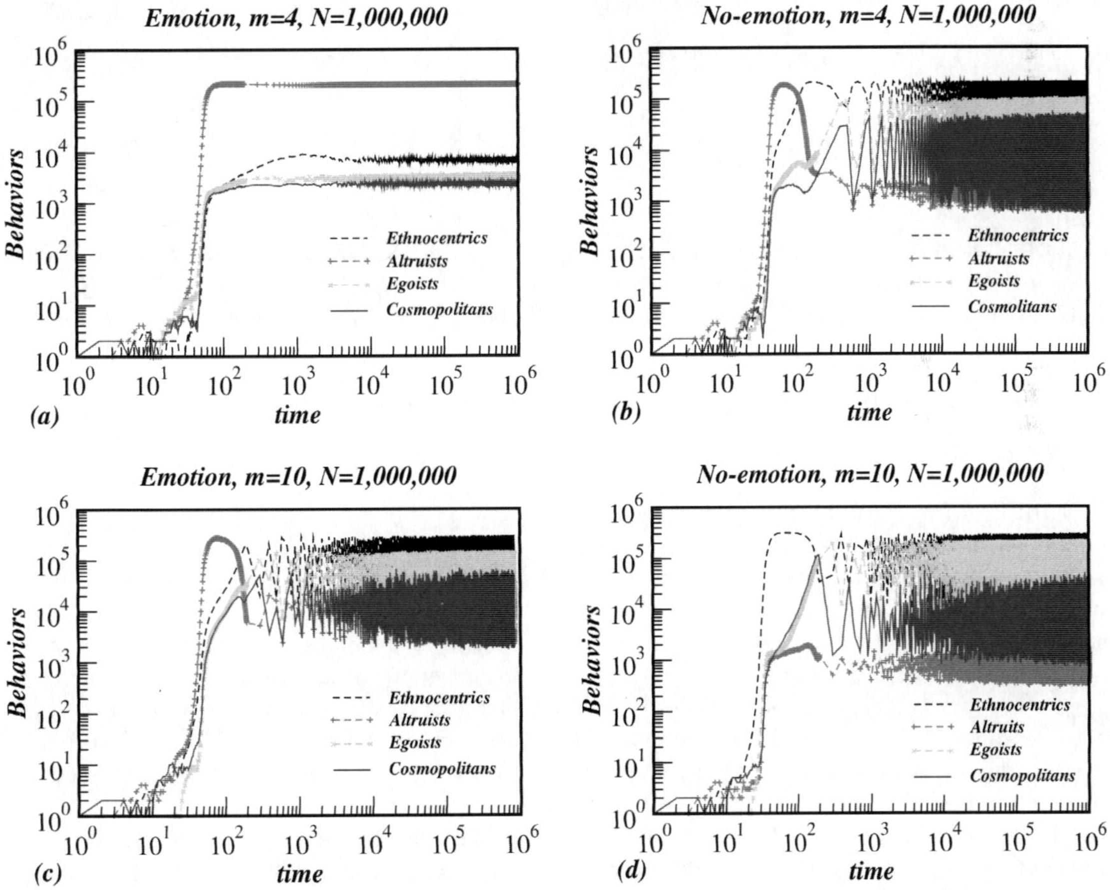

Figure 1. (Color online) Outcomes of the competition of strategies in the modified AH model of tag-based ethnocentrism, showing ethnocentric (black), altruistic (red), egoistic (green), and cosmopolitan (blue) behaviors versus parameter m on ER random graphs networks with emotion (left) and no-emotion (right).

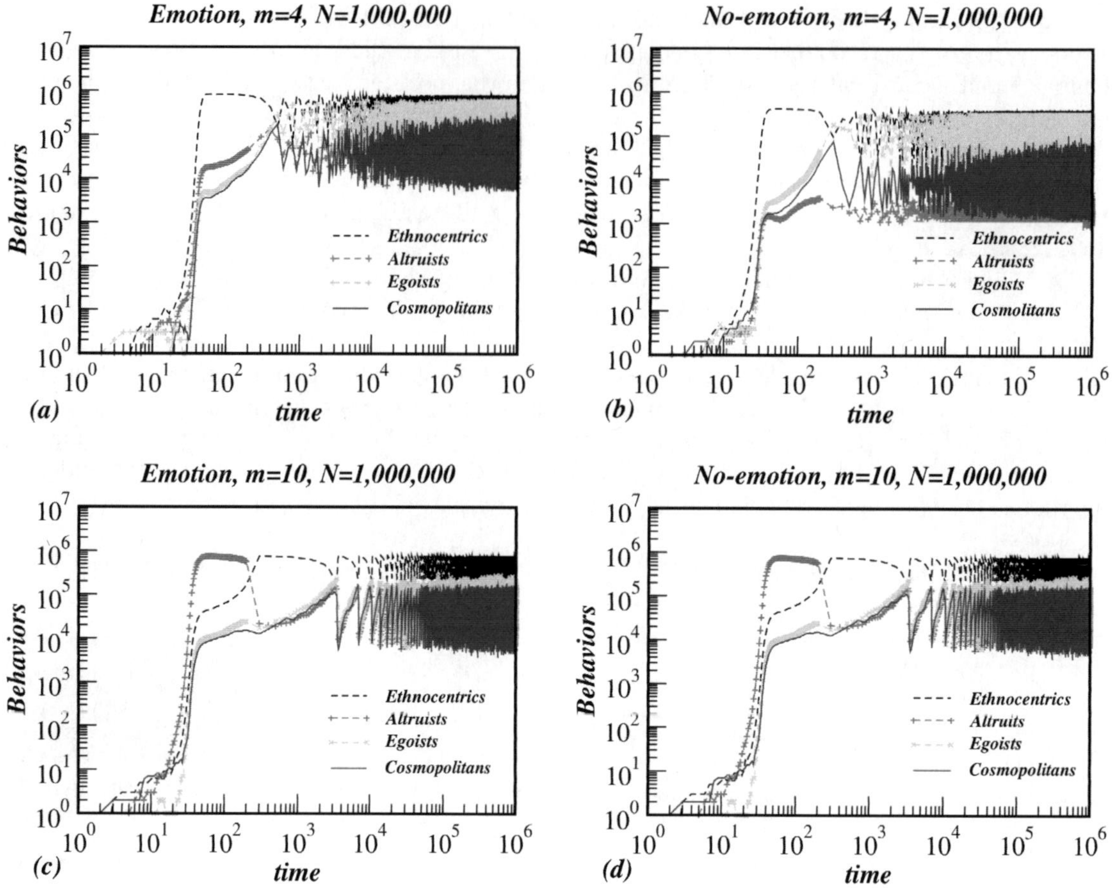

Figure 2. (Color online) As Fig.1, but now on UBA networks.

2. Model and simulations

Our agent-based extension of the classical model of ethnocentrism [5], is simulated here with 1,000,000 nodes (or computational agents) on ER random graphs and UBA networks. We build on our previous model generalizations [35, 36, 37] by studying the effects of random network structure on the evolution of cooperative behavior of asexually reproducing agents with heritable and visually discernible tag and with four different interaction strategies. More specifically, agents are characterized in our model by one out of four possible arbitrary traits called tags, such as different skin colors. We systematically study an evolutionary model on ER random graphs and UBA networks, but with one modification in one of the four distinct steps called Interaction described below [28]. At each iteration, all agents interact pairwise with neighboring co-players and play Prisoner-Dilemma-like games and observe whether a given co-player displays the same or a different tag. Then each agent applies its fixed strategy. Thus, tags and strategies together will influence the final choice, cooperation (help) or defection, of an individual agent. Agents apply one of the four following strategies fixed at birth:

- Ethnocentrism or intra-tag cooperation: Agents help only those with the same tag;
- Cosmopolitanism or extra-tag cooperation: Agents help only those with a different tag;
- Altruism or unconditional cooperation: Agents always help;

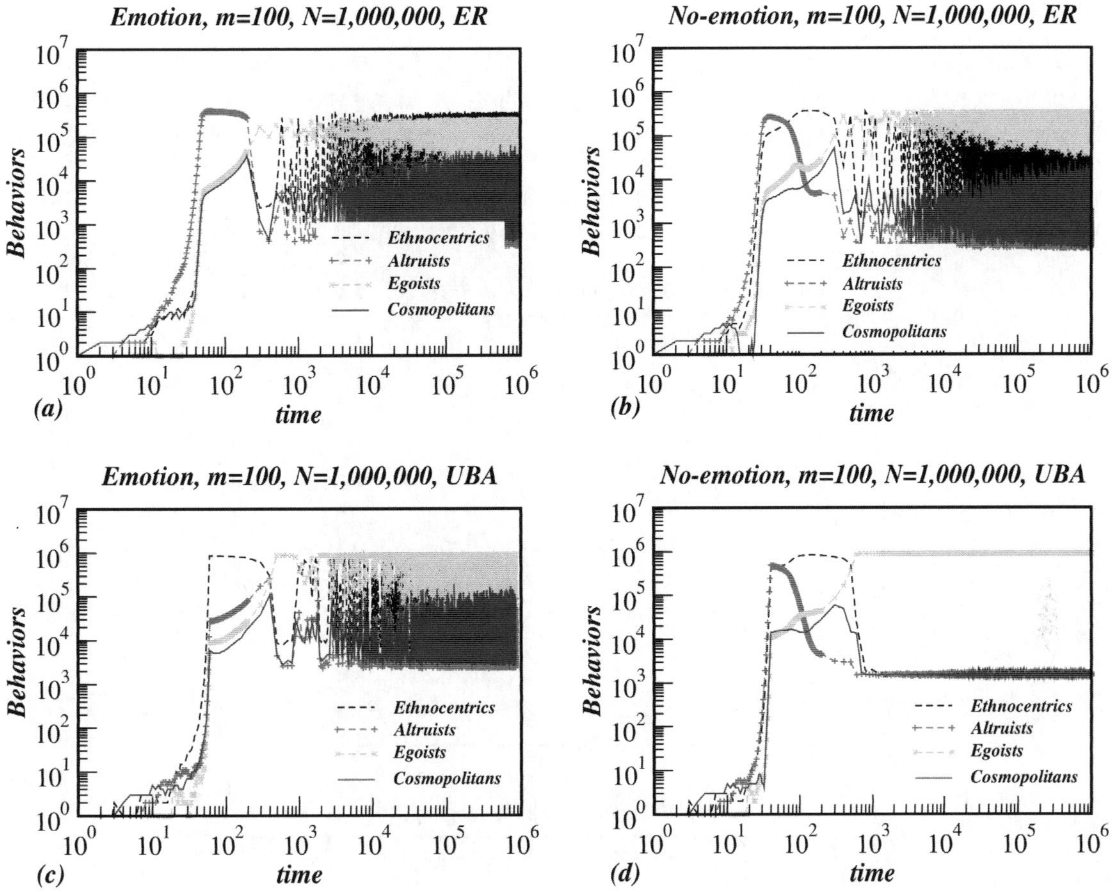

Figure 3. (Color online) Display the behaviors versus time for UER random graphs and UBA networks. Boths for $m = 100$.

- Egoism or unconditional defection: Agents never help.

In our simulations, we started generating an empty network (UBA, ER) which remained static throughout the simulation. Then, an iteration (time step of the simulation) always contained updates consisting of four steps in the following order:

- *1) Immigration.* One immigrant with randomly fixed traits (tag, strategy) lands on a randomly selected empty network node. It gets as reproductive potential α the initial value 0.12.
- *2) Interaction.* Each of N randomly drawn agents searches its connected neighborhood and determines whether to cooperate or to defect, depending on the tag of the neighbor and also considering now the emotion betwen the agents and its neighbors. After donating help, for the donor's α is reduced by the cost $c = 0.01$, while the donation recipient increases its α by the benefit $b = 0.03$. Here, we used a small change where only agents with fitness $\alpha > c$ can help, thus avoiding negative α [28].
- *3) Reproduction.* All agents in regular order check if at least one node in their neighborhood is empty; if yes then with probability α they produce one offspring put on that empty node. The newborns inherit α, as well tag and strategy subject to random mutations with probability $\mu = 0.05$. Thus, reproduction is a simple cloning of the parent.

- *4) Death.* As the final step, each agent is selected to die with probability $\delta = 0.10$.

These four steps constitute one time step or iteration; 10^6 such iterations are made in order to get stationarity. The stabilization procedure of [36] is used for the reproductive potential α: At each time step, all values of α are multiplied by an identical factor such that the average reproductive potential of the population still remains the initial one $\alpha = 0.12$. Finally, for consistency with the original work on ethnocentric cooperation [5] we held this and (most) other parameters constant in accordance with the aforementioned values.

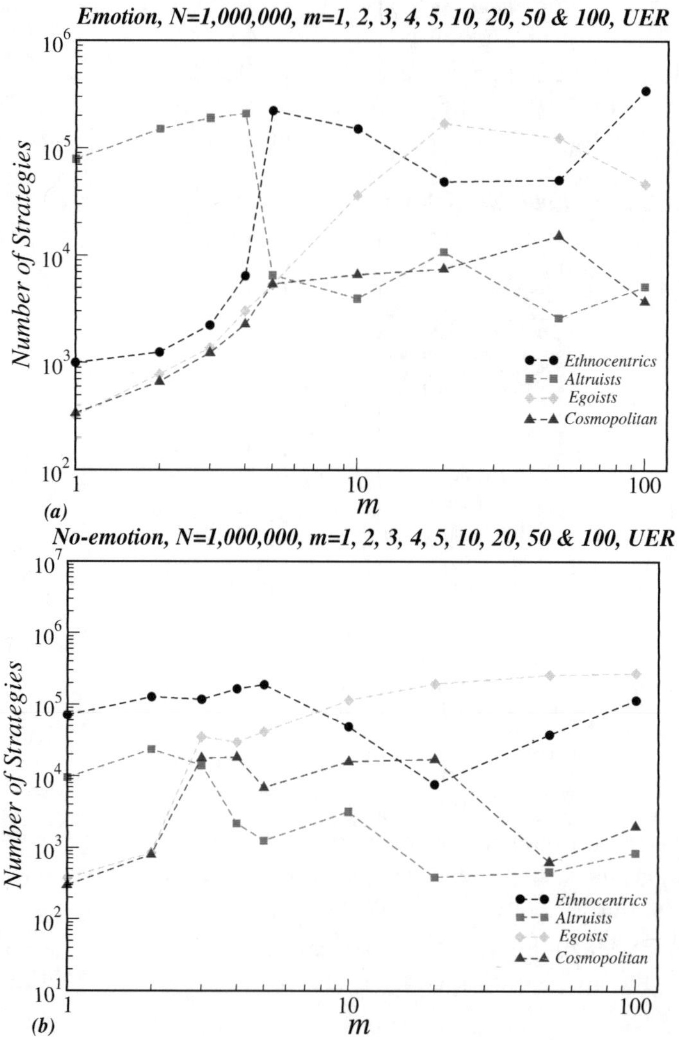

Figure 4. (Color online) Display the number of strategies in our modified AH model of tag-based ethnocentrism, showing ethnocentric (black), altruistic (red), egoistic (green), and cosmopolitan (blue) behaviors versus parameter m on UER .

IOP Conf. Series: Journal of Physics: Conf. Series **1012** (2018) 012009 doi:10.1088/1742-6596/1012/1/012009

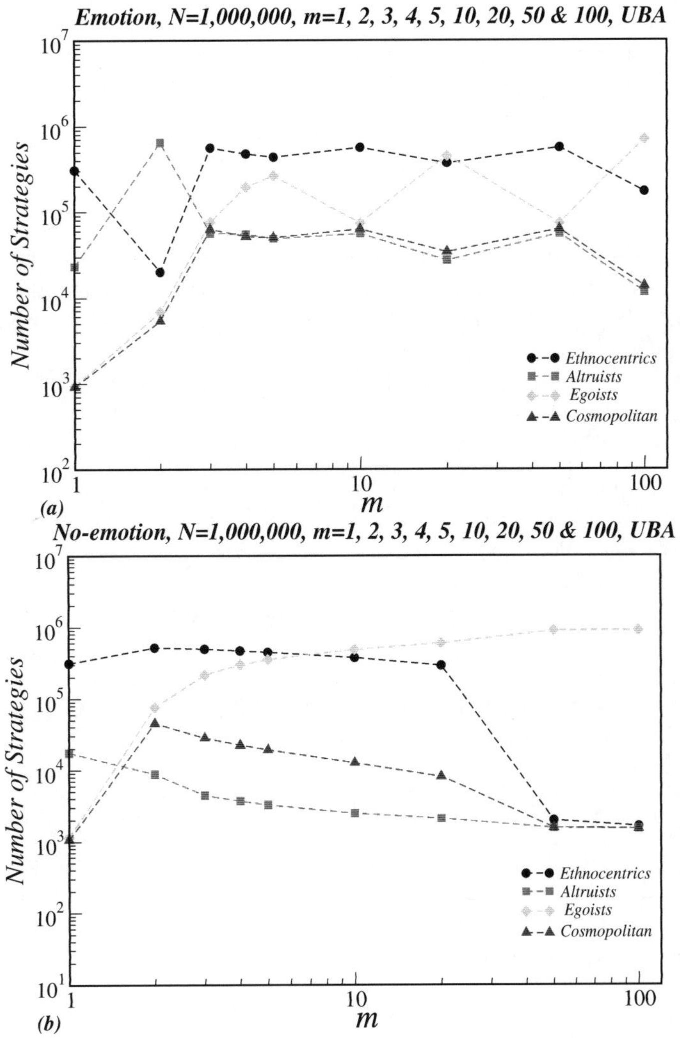

Figure 5. (Color online) As Fig. 4 but now for UBA networks.

3. Results

Figs. 1 and 2 show the numbers of strategists (i.e. their observed behaviors: ethnocentrism, altruism, egoism, and cosmopolitanism) evolving on ER random graphs and UBA networks with parameter $m = 4$ and 10. For $m = 4$ and with emotion the altruism always wins and having ethnocentrism as the second winning strategy, Fig. 1(a). In Fig. 1(b), we observe that ethnocentrism wins after a brief victory in the early stages of altruism. For $m = 10$ and with emotion, Fig. 1(c), we observe a similar behavior of the Fig. 1(b). In the Fig. 1(d) the ethnocentrism always wins and having egoism as the second winning strategy. Fig. 3 compares UBA with ER, both for $m = 100$.

Figs. 4 and 5 show the numbers of strategists for $N = 1,000,000$ agents with $m = 1$, 2, 5, 10, 20, 50 and 100 on UER random graphs and UBA networks for two cases: with emotion and without emotion, respectively. We averaged over the last 500,000 iterations. In Fig. 4(a) already a parameter $m \leq 4$ makes altruism become the winning strategy on UER random graphs and for

$m > 4$ there is intense competition between ethnocentrism and egoism. In Fig. 4(b) for $m \leq 5$ the ethnocentrism wins and for $m > 5$ the egoism is the winning strategy. Fig. 5(a) shows an intense competition between egoism and ethnocentrism. In Fig. 5(b) the ethnocentrism wins for $m \leq 5$ and $m > 5$ the winning strategy is the egoism.

4. Conclusion

Earlier work without emotion [11, 12, 35, 36, 37, 38] on Hammond-Axelrod models [5, 22] using complex networks showed dominance by ethnocentrics and/or altruists, but not by egoists or cosmopolitans. Here we introduced emotion, measuring the fraction of neighbors with the same strategy as the central node. Now, depending on parameters, also egoism can be the winning strategy. An explanation remains to be found.

Acknowledgments

The author is grateful to Dietrich Stauffer for stimulating discussions and for a critical reading of the manuscript. We thank the Conselho Nacional de Desenvolvimento Científico e Tecnológico (CNPq) of Brazil for the financial support and the computational park CENAPAD-UNICAMP-USP, SP-BRAZIL, for their technical support.

References

[1] J. Holland, The effect of labels (Tags) on social interactions, Working Paper 93-10-064, Santa Fe Institute, Sante Fe, New Mexico (1993).
[2] R. L. Riolo, M. D. Cohen and R. Axelrod, Nature 414, 441 (2001).
[3] G. Roberts and T. N. Sherratt, Nature 418, 499 (2002).
[4] R. Axelrod, R. A. Hammond and A. Grafen, Evolution 58, 1833 (2004).
[5] R. A. Hammond and R. Axelrod, J. Confl. Resolut. 50, 926 (2006).
[6] T. Antal, H. Ohtsuki, J. Wakeley, P. D. Taylor and M. A. Nowak, Proc. Natl. Acad. Sci. USA 106, 8597 (2009).
[7] R. A. Laird, J. Theor. Biol. 288, 84 (2011).
[8] R. A. Laird, Int. J. Bifurcation Chaos 22, 1230039 (2012).
[9] E. Cohen, Curr. Anthropol. 53, 588 (2012).
[10] A. Traulsen and H. G. Schuster, Phys. Rev. E 68, 046129 (2003).
[11] F. W. S. Lima, T. Hadzibeganovic and D. Stauffer, Physica A 388, 4999 (2009).
[12] T. Hadzibeganovic, F. W. S. Lima and D. Stauffer, Comput. Phys. Commun. 183, 2315 (2012).
[13] A. Traulsen and J. C. Claussen, Phys. Rev. E 70, 046128 (2004).
[14] A. Traulsen, Eur. Phys. J. B 63, 363 (2008).
[15] F.C. Santos, J.M. Pacheco, T. Lenaerts, Proc. Natl. Acad. Sci. USA, **103**, 3490 (2006).
[16] D.M. Chess, Complex Systems, **2**, 663 (1988).
[17] J. Weibull, Evolutionary Game Theory, MIT University Press, Cambridge, MA, 1996.
[18] S. Wang, M.S. Szalay, C. Zhang, P. Csermely, PLoS ONE, **3** (4), e 1917, (2008).
[19] R. Axelrod, The Evolution of Cooperation, Basic Books, New York, 1984.
[20] C. Castellano, S. Fortunato, V. Loreto, Rev.Modern Phys. **81**, 591 (2009).
[21] F.C. Santos, J.M. Pacheco, J. Evolution Biol. **19**, 726 (2006).
[22] R.A. Hammond, R. Axelrod, Theor. Popul. Biol. **69**, 333 (2006).
[23] M. Schnegg, Internat. J. Modern Phys. C **17**, 1067 (2006).
[24] R.N. Costa-Filho, M.P. Almeida, J.S. Andrade Jr., J.E. Moreira, Phys. Rev. E 60, 1067 (1999) .
[25] J. Geanakoplos *et al.*, American Economic Review 102, 53 (2012).
[26] L.A.N. Amaral, A. Scala, M. Barthelemy, H.E. Stanley, Proc. Natl. Acad. Sci. USA **97**, 11149 (1999).
[27] S.N. Dorogovtsev, J.F.F. Mendes, Evolution of Networks: From Biological Nets to the Internet and WWW, Oxford Univ. Press, Oxford, UK, 2003.
[28] F. W. S. Lima, Int. J. Mod. Phys. C **26**, 1550135 (2015).
[29] R. Albert, A.-L. Barabási, Rev. Modern Phys. **74**, 47 (2002).
[30] M.A. Sumour, M.A. Radwan, Int. J. Mod. Phys. C **23**, 1250062 (2012).
[31] A. Aleksiejuk, J.A. Hołyst, D. Stauffer, Physica A **310**, 260 (2002).
[32] R. Heinsohn, C. Packer, Science **269**, 1260 (1995).
[33] M.A. Sumour, M.M. Shabat, Int. J. Mod. Phys. C **16**, 585 (2005).

[34] D. Stauffer, S. Moss de Oliveira, P.M.C de Oliveira and J.S.Sá Martins, *Biology, Sociology, Geology by Computational Physicists.* Elsevier, Amsterdam (2006).
[35] F. W. S. Lima, T. Hadzibeganovic and D. Stauffer, Int. J. Mod. Phys. C **25**, 1450006 (2014).
[36] T. Hadzibeganovic, F. W. S. Lima and D. Stauffer, Behavioral Ecology and Sociobiology **68**, 1059 (2014).
[37] F. W. S. Lima, T. Hadzibeganovic and D. Stauffer, Physica A **388**, 4999 (2009).
[38] T. Hadzibeganovic, D. Stauffer, and Xiao-Pu Han. Behavioral Processes **113**, 86 (2015).

IOP Conf. Series: Journal of Physics: Conf. Series **1012** (2018) 012010 doi:10.1088/1742-6596/1012/1/012010

Filament instability under constant loads

A. G. Monastra[1,4,*], **M. F. Carusela**[1,4], **M. V. D'Angelo**[2,4] **and L. Bruno**[3,4]

[1] Instituto de Ciencias, Universidad Nacional de Gral. Sarmiento, Los Polvorines, Buenos Aires, Argentina

[2] Grupo de Medios Porosos, Facultad de Ingeniería , Universidad Nacional de Buenos Aires, Argentina

[3] Departamento de Física & IFIBA-CONICET, Facultad de Ciencias Exactas y Naturales, Universidad Nacional de Buenos Aires, Argentina

[4] Consejo Nacional de Investigaciones Científicas y Técnicas, Argentina

E-mail: *amonast@ungs.edu.ar

Abstract. Buckling of semi-flexible filaments appears in different systems and scales. Some examples are: fibers in geophysical applications, microtubules in the cytoplasm of eukaryotic cells and deformation of polymers freely suspended in a flow. In these examples, instabilities arise when a system's parameter exceeds a critical value, being the Euler force the most known. However, the complete time evolution and wavelength of buckling processes are not fully understood. In this work we solve analytically the time evolution of a filament under a constant compressive force in the small amplitude approximation. This gives an insight into the variable force scenario in terms of normal modes. The evolution is highly sensitive to the initial configuration and to the magnitude of the compressive load. This model can be a suitable approach to many different real situations.

1. Introduction

The beam theory stablishes that when a compressive force is applied to the end of a slender rod, it collapses and buckles when the force exceeds a critical value [1, 2]. The shape of the buckled filament depends on the boundary conditions. The Euler force depends on the rod geometry and its elasticity. For a clamped-clamped filament its value is:

$$P_c = 4\pi^2 \frac{EI}{L^2} \ , \qquad (1)$$

where E, I and L are the filament's Young modulus, the second moment of inertia and length, respectively. Importantly, the critical force does not depend on the environment where the buckling takes place. However, the way the buckling proceeds after the critical force is surpassed (e.g. the characteristic time taken for the filament to achieve a stationary shape) does strongly depend on the surrounding viscosity [3]. In Fig. 1 a slender filament immersed in glycerol compressed under two different loads is shown. For forces slightly larger that the Euler one the filament displays a sinusoidal shape with a characteristic wavelength of the order of its length. On the other hand, if the load largely exceeds the critical value, the filament deforms more rapidly and with a shorter wavelength.

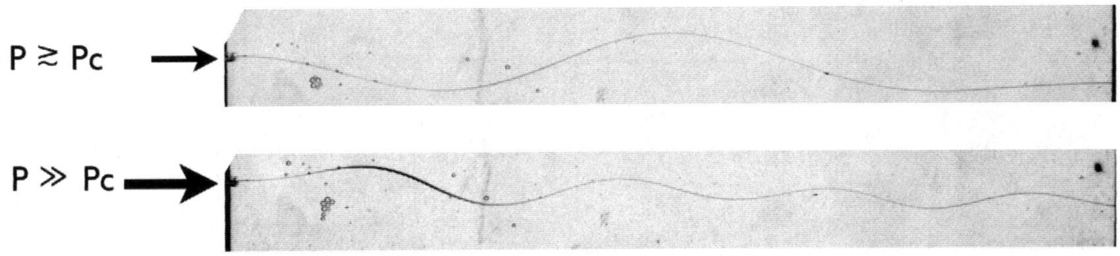

Figure 1. Snapshots of a compressing polyethylene terephthalate filament in glycerol for two different values of the applied force. The length of the filament is 30 cm.

Deformation of flexible fibers is of interest in areas such as physics, biology or engineering, and in view of their potential applications in many different industrial fields. Examples include the manufacturing of fiber-reinforced composites [4, 5], the rheology of biological polymers [6], and the motility of microorganisms [7, 8]. The use of long optical fibers has been suggested as a method to realize distributed in situ measurements (temperature for instance) on natural water flows [9]. Also fibers are widely used by petroleum engineers [10] to enhance the proppant transport capabilities of fracturing fluids, or to prevent the backflow of proppant. Other domains of application are civil engineering (special cements, structural reinforcement), textile engineering, bio engineering and medicine.

In this work we develop an general description of the behaviour of filaments immersed in a viscous medium under compression in terms of normal modes. Our approach intends to explore a general phenomenon that can be found in a variety of micro to macro scales systems and applications.

2. Theoretical approach

The shape of a semi-flexible filament is determined by the position of its neutral axis $\mathbf{r}(l)$, with l a curvilinear coordinate along the filament varying from 0 to L. This configuration has an elastic energy due to strain ε given by:

$$V_{\mathrm{E}} = \frac{1}{2}EA\int_0^L \varepsilon^2(l)\mathrm{d}l = \frac{1}{2}EA\int_0^L (|\mathbf{r}'(l)| - 1)^2\mathrm{d}l \ , \tag{2}$$

where A is the transversal area of the filament. Also, the filament has a bending energy given by its curvature $\mathcal{C}$:

$$V_{\mathrm{B}} = \frac{1}{2}EI\int_0^L \mathcal{C}^2(l)(1+\varepsilon(l))^2\mathrm{d}l = \frac{1}{2}EI\int_0^L \frac{|\mathbf{r}'(l)\times\mathbf{r}''(l)|^2}{|\mathbf{r}'(l)|^4}\mathrm{d}l \ . \tag{3}$$

In both formulas the primes indicate derivative with respect to the coordinate l. Applying a Lagrangian minimization principle, one can compute the forces on an infinitesimal segment $\mathrm{d}l$ at position l due to strain (elastic force $\mathbf{f}_{\mathrm{E}}$) and curvature (bending force $\mathbf{f}_{\mathrm{B}}$), which are produced by the neighbouring segments. The full expressions are rather complicate, involving up to the fourth derivative of the position $\mathbf{r}(l)$.

Since the filament is immersed in an homogeneous viscous medium, we consider the viscous drag on a filament element as $\mathbf{f}_{\mathrm{V}} = -c\dot{\mathbf{r}}$. In this expression c is the drag coefficient per unit length, which is proportional to the dynamical viscosity, and the dot accounts for time derivative.

In an overdamped regime, the inertia term can be neglected and the equation of motion for an infinitesimal segment is given by:

$$\mathbf{f}_{\rm E} + \mathbf{f}_{\rm B} + \mathbf{f}_{\rm V} = 0 \ . \tag{4}$$

If the filament motion is constrained to two dimensions, its shape can be parametrized as:

$$\mathbf{r}(l) = (l + \delta x(l); \delta y(l); 0) \ , \tag{5}$$

and we consider a small amplitude approximation $\delta x, \delta y \ll L$.

Replacing this parametrization in Eq. (4), it can be shown that $\delta x''$ and higher order derivatives are proportional to second order derivatives of δy. Moreover, the time derivative $\delta \dot{x}$ is exponentially small after a short initial time $\tau_{\rm E}$ given by

$$\tau_{\rm E} = \frac{c}{EA}\left(\frac{L}{\pi}\right)^2 \ . \tag{6}$$

Also, when the filament is under a compressive load P, at first order we obtain for the strain

$$\delta x'(l) = -\frac{P}{EA} \ , \tag{7}$$

which is usually much smaller than one. Taking into account this result, the amplitude in the transversal direction δy is determined by the following equation:

$$\left(1 + \frac{2P}{EA}\right) EI\delta y'''' + P\delta y'' + c_{\perp}\delta \dot{y} = 0 \ . \tag{8}$$

This is the so called hydrodinamic equation [3]. To solve this linear differential equation, we propose separation of variables and exponential dependence:

$$\delta y(l,t) = \mathcal{C}\exp(kl)\exp(\Gamma t) \ . \tag{9}$$

Replacing this proposed solution into Eq. (8) we obtain

$$\left(1 + \frac{2P}{EA}\right) EIk^4 + Pk^2 + c\Gamma = 0 \ . \tag{10}$$

This is a fourth order polynomial in k which gives four possible solutions which can be superposed. It is convenient to redefine the parameters as:

$$p = \frac{L^2 P}{\left(1 + \frac{2P}{EA}\right) EI} \ , \tag{11}$$

$$\gamma = \frac{cL^4\Gamma}{\left(1 + \frac{2P}{EA}\right) EI} = \tau_{\rm B}\Gamma \ , \tag{12}$$

$$\kappa = kL \ . \tag{13}$$

The value p corresponds to a dimensionless compressing load and γ is the dimensionless decaying/growing rate, whose natural scale is a characteristic *bending* time $\tau_{\rm B}$, which is much bigger than $\tau_{\rm E}$ for a slender filament. In terms of these dimensionless parameters the characteristic polynomial can be written as:

$$\kappa^4 + p\kappa^2 + \gamma = 0 \ . \tag{14}$$

For $\gamma < 0$, which corresponds to a decreasing solution in time, the four possible solutions are $\kappa = \{\mathrm{i}\kappa_1; -\mathrm{i}\kappa_1; \kappa_2; -\kappa_2\}$, with

$$\kappa_1 = \sqrt{\frac{\sqrt{p^2 - 4\gamma} + p}{2}} \,, \tag{15}$$

$$\kappa_2 = \sqrt{\frac{\sqrt{p^2 - 4\gamma} - p}{2}} \,. \tag{16}$$

Therefore in this case the general solution can be written as a superposition of circular and hyperbolic trigonometric functions.

For $0 < \gamma < p^2/4$, the amplitude grows exponentially, and there are four possible pure imaginary solutions $\kappa = \{\mathrm{i}\kappa_1; -\mathrm{i}\kappa_1; \mathrm{i}\kappa_3; -\mathrm{i}\kappa_3\}$, with κ_1 as in Eq. (15), and

$$\kappa_3 = \sqrt{\frac{p - \sqrt{p^2 - 4\gamma}}{2}} \,. \tag{17}$$

These solutions can be written as circular trigonometric solutions. For $\gamma > p^2/4$ the four possible values of κ are complex. Nevertheless, it turns out that these solutions do not arise for the boundary conditions that we study. The general solutions can be written as

$$\delta y(l,t) = [\mathcal{C}_1 \cos(\kappa_1 l/L) + \mathcal{D}_1 \sin(\kappa_1 l/L) + \mathcal{C}_2 \cosh(\kappa_2 l/L) + \mathcal{D}_2 \sinh(\kappa_2 l/L)] \exp(\gamma t/\tau_{\mathrm{B}}), \tag{18}$$

$$\delta y(l,t) = [\mathcal{C}_1 \cos(\kappa_1 l/L) + \mathcal{D}_1 \sin(\kappa_1 l/L) + \mathcal{C}_3 \cos(\kappa_3 l/L) + \mathcal{D}_3 \sin(\kappa_3 l/L)] \exp(\gamma t/\tau_{\mathrm{B}}), \tag{19}$$

for negative and positive γ, respectively. The constants $\mathcal{C}_i$ and $\mathcal{D}_i$ should be much smaller than L to fulfill the small deformation approximation, and in the case of a growing solution with $\gamma > 0$, it will only be valid for a time such that $\mathcal{C}_i \exp(\gamma t/\tau_{\mathrm{B}}) \ll L$.

We are particularly interested in the clamped-clamped boundary condition, as shown in Fig. 1, where both ends remain horizontal, and always at the same vertical position. Due to the symmetry of these boundary conditions, it is more convenient to shift the coordinate l, going from $-L/2$ to $L/2$. Therefore the clamped-clamped boundary condition corresponds to $\delta y(\pm L/2, t) = 0$ and $\delta y'(\pm L/2, t) = 0$. These four conditions provide us four equations to determine the constants $\mathcal{C}_i$ and $\mathcal{D}_i$. Moreover, with simple sums and substractions of the equations, they decouple to

$$\mathcal{C}_1 \cos(\kappa_1/2) + \mathcal{C}_2 \cosh(\kappa_2/2) = 0 \,, \tag{20}$$

$$-\mathcal{C}_1\kappa_1 \sin(\kappa_1/2) + \mathcal{C}_2\kappa_2 \sinh(\kappa_2/2) = 0 \,, \tag{21}$$

$$\mathcal{D}_1 \sin(\kappa_1/2) + \mathcal{D}_2 \sinh(\kappa_2/2) = 0 \,, \tag{22}$$

$$\mathcal{D}_1\kappa_1 \cos(\kappa_1/2) + \mathcal{D}_2\kappa_2 \cosh(\kappa_2/2) = 0 \,. \tag{23}$$

The first two equations determine symmetric solutions, while the last two antisymmetric ones. In order to avoid the trivial solution, the determinant for each pair of equations should be zero. For $\gamma < 0$:

$$\mathcal{J}^{(+)}(p,\gamma) = \kappa_2 \sinh(\kappa_2/2) \cos(\kappa_1/2) + \kappa_1 \sin(\kappa_1/2) \cosh(\kappa_2/2) \,, \tag{24}$$

$$\mathcal{J}^{(-)}(p,\gamma) = \kappa_2 \cosh(\kappa_2/2) \sin(\kappa_1/2) - \kappa_1 \cos(\kappa_1/2) \sinh(\kappa_2/2) \,. \tag{25}$$

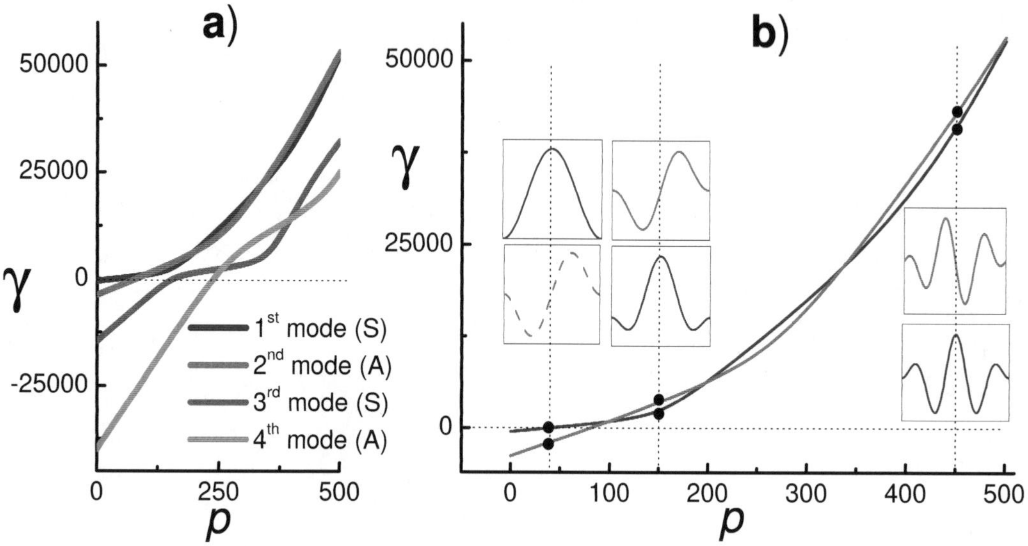

Figure 2. (a) Plot of dimensionless characteristic rates γ versus compressive force p for the first symmetric (S) and antisymmetric (A) modes. (b) Detail for the first symmetric and antisymmetric modes. The insets display the filament shapes for $p = 40, 150$ and 450. Dashed line represents a mode with a decaying amplitude.

We can proceed in the same way for the case $0 < \gamma < p^2/4$, and the determinants are:

$$\mathcal{J}^{(+)}(p,\gamma) = \kappa_3 \sin(\kappa_3/2)\cos(\kappa_1/2) - \kappa_1 \sin(\kappa_1/2)\cos(\kappa_2/2) \ , \quad (26)$$

$$\mathcal{J}^{(-)}(p,\gamma) = \kappa_3 \cos(\kappa_3/2)\sin(\kappa_1/2) - \kappa_1 \cos(\kappa_1/2)\sin(\kappa_3/2) \ . \quad (27)$$

The dimensionless parameter p is determined using Eq. (11). The values of γ that vanish the corresponding determinants, provide *quantized* hydrodynamic modes.

We plot these quantized values of γ as a function of the parameter p in Fig. 2 for the first four modes (those with the highest value of γ). For $p < 4\pi^2$ all modes have a negative value of γ, corresponding to a compressive force smaller than the critical value. As this value is surpassed, there are growing modes which contribute to the buckling process, being the first two the fastest ones. We also observe an interesting successive crossovers between the symmetric and antisymmetric modes. This implies that for some range of compressing force p, the antisymmetric mode grows faster than the symmetric one, and for other ranges the opposite situation is observed (see Fig. 2).

The shape of the different modes, for decaying rates $\gamma < 0$ are:

$$\Psi_n^{(+)}(p,u) = \mathcal{K}_n \left[\cosh(\kappa_2/2)\cos(\kappa_1 u) - \cos(\kappa_1/2)\cosh(\kappa_2 u)\right] \ , \quad (28)$$

$$\Psi_n^{(-)}(p,u) = \mathcal{K}_n \left[\sinh(\kappa_2/2)\sin(\kappa_1 u) - \sin(\kappa_1/2)\sinh(\kappa_2 u)\right] \ , \quad (29)$$

corresponding to symmetric and antisymmetric modes, respectively. The dimensionless coordinate along the filament $u = l/L$ ranges from -1/2 to 1/2. For growing rates $\gamma > 0$ the filament shapes are:

$$\Psi_n^{(+)}(p,u) = \mathcal{K}_n\left[\cos(\kappa_3/2)\cos(\kappa_1 u) - \cos(\kappa_1/2)\cos(\kappa_3 u)\right] , \tag{30}$$
$$\Psi_n^{(-)}(p,u) = \mathcal{K}_n\left[\sin(\kappa_3/2)\sin(\kappa_1 u) - \sin(\kappa_1/2)\sin(\kappa_3 u)\right] . \tag{31}$$

These modes depend parametrically on the compressive force. For a given value of p, the different possible discretized values of γ determine the corresponding values of κ_1 and κ_2 (or κ_3 if $\gamma > 0$) through Eqs. (15), (16) and (17). For a convenient choice of the arbitrary constants $\mathcal{K}_n$, the different modes are orthonormal, fulfilling

$$\int_{-1/2}^{1/2} \Psi_n(p,u)\Psi_m(p,u)\mathrm{d}u = \delta_{nm} \tag{32}$$

The general solution for a constant compressing load P is

$$\delta y(l,t) = \sum_{n=1}^{\infty} a_n \Psi_n(p, l/L)\exp(\gamma_n t/\tau_\mathrm{B}) \tag{33}$$

where the coefficients a_n are determined from the initial condition of the filament projected on each eigenmode. As an example, in Fig. 2 (b), we plot the shape of the first two modes. We observe that as the parameter p increases, the modes have more nodes and the typical wavelengths are shorter. This situation describes the experimental behaviour observed in Fig. 1.

3. Conclusions

In this paper we have explored the buckling dynamics of a semi-flexible filament immersed in a viscous environment. We have found analytical solutions of the hydrodynamics beam equation in the small deformation limit in terms of hydrodynamics modes. These modes have characteristic relaxation times and critical forces, which depend on geometrical and mechanical parameters of the filament, the medium viscosity and the magnitude of the compression. When the compression load is very small, the filament shape straightens. However, if the compression magnitude surpassed a critical value, an interesting scenario occurs: the buckling shape will depend on the magnitude of the compression. Moreover, successive crossovers between symmetric and antisymmetric modes are observed, depending on the particular value of the compressive load.

These results show that the shape of a filament under compression is more complex than that of a single sinusoidal mode. This work provides a theoretical explanation that can help to understand the evolution of filament buckling when a variable force is applied, which it is the case in most experiments. The spatio-temporal characterization of the buckling can have significant importance in many field applications where a flexible filament is dramatically compressed within a viscous environment.

Acknowledgments

All authors are members of CIC-CONICET. AGM and MFC want to thank for the funding project PIO-CONICET-UNGS.

References

[1] Landau L D and Lifshitz E M 1986 *Theory of Elasticity*, vol 7, Course of Theoretical Physics (New York: Elsevier)
[2] Timoshenko S P and Gere J M 1963 *Theory of Elastic Stability* (New York: McGraw-Hill Book Company)
[3] Howard J 2001 *Mechanics of motor proteins and the cytoskeleton* (Sunderland, MA: Sinauer Associates, Publishers)
[4] Yasuda K, Mori N and Nakamura K 2002 A new visualization technique for short fibers in a slit flow of fiber suspensions *International Journal of Engineering Science* **40** 1037-1052

[5] Yoo DY, Kim S, Park GJ, Park JJ and Kim SW 2017 Effects of fiber shape, aspect ratio, and volume fraction on flexural behavior of ultra-high-performance fiber-reinforced cement composites *Composite Structures* **174** 375-388
[6] Lagomarsino M, Pagonabarraga I and Lowe C 2005 Hydrodynamic induced deformation and orientation of a microscopic elastic filament *Phys. Rev. Lett.* **94** 148104
[7] Lowe C 2003 Dynamics of filaments: modelling the dynamics of driven microfilaments *Philos. Trans. R. Soc. London* B **358** 1543-1550
[8] Chaban B, Velocity Hughes H and Beeby M 2015 The flagellum in bacterial pathogens: For motility and a whole lot more *Seminars in Cell & Developmental Biology* **46** 91-103
[9] Selker J, van de Giesen M, Westhoff N, Luxemburg W and Parlange MB 2006 Fiber optics opens window on stream dynamics *Geophys. Res. Lett.* **33** L24401
[10] Howard PR, King MT, Morris M, Feraud JP, Slusher G and Lipari S 1995 Fiber/proppant mixtures control proppant flowback in South Texas *SPE Annual Technical Conference and Exhibition* (Dallas, TX: Society of Petroleum Engineers) pp 453-454

Brownian Dynamics simulations of model colloids in channel geometries and external fields

Ullrich Siems, Peter Nielaba

Fachbereich Physik, University of Konstanz, 78457 Konstanz, Germany

E-mail: ullrich.siems@uni-konstanz.de

Abstract. We review the results of Brownian Dynamics simulations of colloidal particles in external fields confined in channels. Super-paramagnetic Brownian particles are well suited two-dimensional model systems for a variety of problems on different length scales, ranging from pedestrian walking through a bottleneck to ions passing ion-channels in living cells. In such systems confinement into channels can have a great influence on the diffusion and transport properties. Especially we will discuss the crossover from single file diffusion in a narrow channel to the diffusion in the extended two-dimensional system. Therefore a new algorithm for computing the mean square displacement (MSD) on logarithmic time scales is presented. In a different study interacting colloidal particles were dragged over a washboard potential and are additionally confined in a two-dimensional micro-channel. In this system kink and anti-kink solitons determine the depinning process of the particles from the periodic potential.

1. Introduction

Colloidal particles are a well suited model system [1] to study principle phenomena on all length scales ranging from gravitational collapse [2], over pedestrian walking [3], to two dimensional crystallization [4, 5, 6], glass transition [7] and non-equilibrium phase transition [8]. Besides the use of colloids as model systems the knowledge of their dynamics is crucial for the development of microscopic lab-on-the-chip devices [9] or devices for controlled drug release [10].

This proceeding presents some interesting results of computer simulations of such colloidal model systems. We use Brownian dynamics simulation [11], which is an Euler integration of the overdamped Langevin equation. The positions $\mathbf{r}_i(t)$ of particle i are updated to a new position after the time step Δt

$$\mathbf{r}_i(t+\Delta t) = \mathbf{r}_i(t) + \frac{D_0 \Delta t}{k_B T}\mathbf{F}_i + \sqrt{2D_0\Delta t}\mathbf{R}(t) \ , \qquad (1)$$

with microscopic diffusion constant D_0, thermal energy $k_B T$, delta-correlated Gaussian random numbers $\mathbf{R}(t)$ with zero-mean and unit variance, and the deterministic force $\mathbf{F}_i$ acting on this particle. The particles are restricted to the xy-plane, and confined in a channel. Periodic boundary condition is applied in x-direction and hard-wall boundary condition is applied in y-directions. The hard-wall boundary is realized by using the analytically known transition probability of a Brownian particle near a hard boundary as proposed by Behringer and Eichhorn [12]. The length L_X of the channel in x-direction is much larger than the width L_Y in y-directions to avoid finite size effects.

The particles interact via the repulsive potential of parallel dipoles $V = \epsilon(\sigma/r_{ij})^3$, with interaction strength ϵ, particle diameter σ and inter-particle distance $r_{ij} = |\mathbf{r}_i - \mathbf{r}_j|$. This models the interaction of super-paramagnetic colloidal particles in a perpendicular magnetic field, as they were used for experiments in two dimensions [6, 13, 14].
The liquid-solid phase transition in this system is described by the dimensionless control parameter [15]

$$\Gamma = \frac{\epsilon(\sigma)^3}{k_B T R^3} \tag{2}$$

which is given by the ratio of potential energy to thermal energy $k_B T$.

2. Diffusion in channels

In the restricted geometry the particles form layers parallel to the wall [15, 16]. The layering is favored, if the width L_Y of the channel and the spacing R between the layers are commensurable, and otherwise suppressed. In [17] we studied the diffusion in channels with multiple layers of particles at an interaction strength of $\Gamma \approx 12$ near disorder to layer transition. As can be seen from the particle trajectory (Fig. 1 b)) the diffusion strongly depends on the width L_Y of the channel. If layering is favored, the particles diffuse for a long time on their own layer, and only for even longer times they can switch between the layers. If layering is suppressed, the particles can freely diffuse throughout the channel, as in a liquid.
The mean square displacement can be approximated on an intermediate time scale by a power

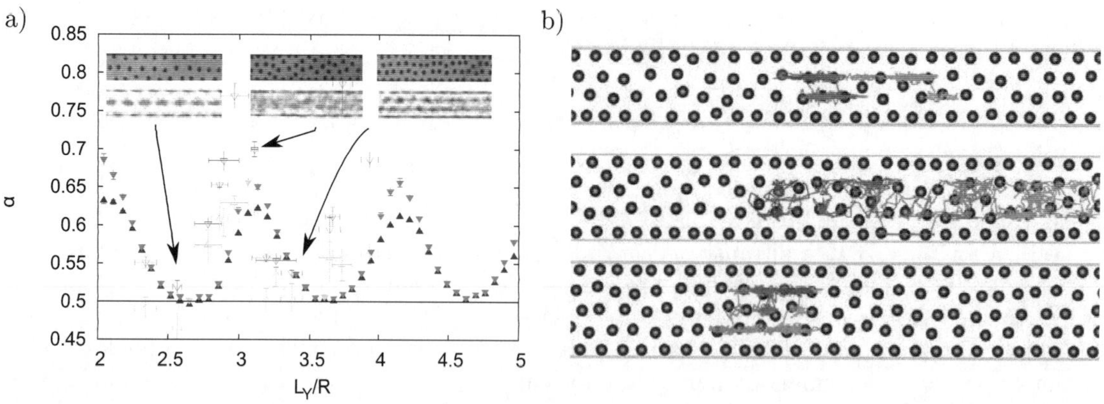

Figure 1. a) Minimal diffusion exponent as a function of reduced channel width L_Y/R, comparing experiments (open symbols with errorbars) and simulation (closed symbols). b) Snapshots of channels of with four layers, five layers and one in between, as well as the trajectory of one particle in each channel with time coloring from green to red. (See Ref. [17])

law $\langle \Delta x^2(t) \rangle \propto t^\alpha$ with exponent α. This exponent is a more universal quantity. For a free particle one finds the normal diffusion exponent $\alpha = 1$ and for particles in a small channel where mutual passing is forbidden one finds the single-file diffusion exponent $\alpha = 0.5$.
This intermediate diffusion exponent is plotted in Figure 1 a) as a function of dimensionless channel width L_Y/R. We see an oscillatory dependency for the simulation results, which is due to the layer structure. If the particles are ordered in layers, the single-file exponent 0.5 can be observed in the intermediate time regime. Otherwise, the MSD is still sub-diffusive in this time regime, but with an exponent greater than the single-file exponent 0.5.

3. Mean Square Displacement

The mean square displacement (MSD) is a quantity that is often used to describe the time dependent self-diffusion of colloidal particles. It measures the averaged squared distance Δx^2 which a particle travels in a time interval Δt. This quantity is averaged over all N particles in the system, and over all possible time origins t_0. Therefore the MSD is calculated by

$$\langle \Delta x^2(\Delta t) \rangle = \frac{1}{N \cdot N_{t_0}} \sum_{t_0} \sum_{i=1}^{N} (x_i(t_0 + \Delta t) - x_i(t_0))^2 \ . \tag{3}$$

In a finite simulation of length t_{max} the time differences Δt and the time origins t_0 can be integral multiples of the time step δt and must obey $t_0 + \Delta t \leq t_{max}$. The calculation of the MSD is in this sense of the order of $O(t_{max}^2)$ and therefore becomes the bottleneck of each simulation that normally has a linear time scaling. For this calculation also all positions have to be stored over the total simulation time.

This effort can be reduced drastically: First, because for large time differences Δt the use of each time step as a new time origin t_0 results in an average over multiple dependent time windows. Second, because the MSD is a quantity which varies fast at small Δt and slow at larger Δt. One usually depicts the MSD in a log-log-plot. Hence less sampling points are needed for larger time differences than for smaller ones.

The so called order-n algorithm by Frenkel and Smit [18] solves both of theses problems. But the algorithm is complex, not applicable to other correlation function than the MSD and the sampling points are not evenly spaced on a logarithmic scale. A detailed description of this algorithm is given in the paper of Dubbeldam *et al.* [19]. They also provide an extension of the order-n algorithm to all correlation functions. We present here another method to calculate the MSD much simpler and even a little bit more efficient.

An easier method to reach higher sampling frequencies at small Δt and lower sampling frequencies at large Δt is to directly define the sampling points Δt_n at which the MSD should be calculated. The sampling points should be equally spaced on a logarithmic scale and integral multiples of the time step δt. For simplicity we set the time step $\delta t = 1$. Then the condition for equally spaced integers on a logarithmic scale would be fulfilled for example for sampling points $\Delta t_n = 2^n$ at powers of two. But this would lead to only three points per decade. We choose a power law a^n with a smaller base $2 > a > 1$. For a value of $a = 10^{\frac{1}{15}} \approx 1.166$ we achieve 15 sampling points per decade. To fulfill the condition of integral values we define the time differences

$$\Delta t_n = \mathrm{nint}(a^{n+1-n_{min}} - a^{n_{min}}) \ , \tag{4}$$

with a rounding function nint() and a shift by the constant $n_{min} = \mathrm{nint}\left(1 - \frac{\log(a-1)}{\log(a)}\right)$ which keeps the function injective.

At first glance one could think that it would be sufficient to sample only over non overlapping intervals. Which means to set only every Δt_n a new time origin t_0 for the calculation of $\langle \Delta x^2(\Delta t_n) \rangle$. But this is not true. To achieve a better accuracy than the order-n algorithm one defines a sampling rate

$$R_n = \mathrm{nint}\left(\Delta t_n^b + \frac{1}{2}\right) \ , \tag{5}$$

with constant $b = \left(\frac{\log(30)}{\log(t_{max})}\right)$ for which every R_n time steps a new time origin t_0 is set. The algorithm computes the mean square displacement at the sampling points Δt_n by summing over

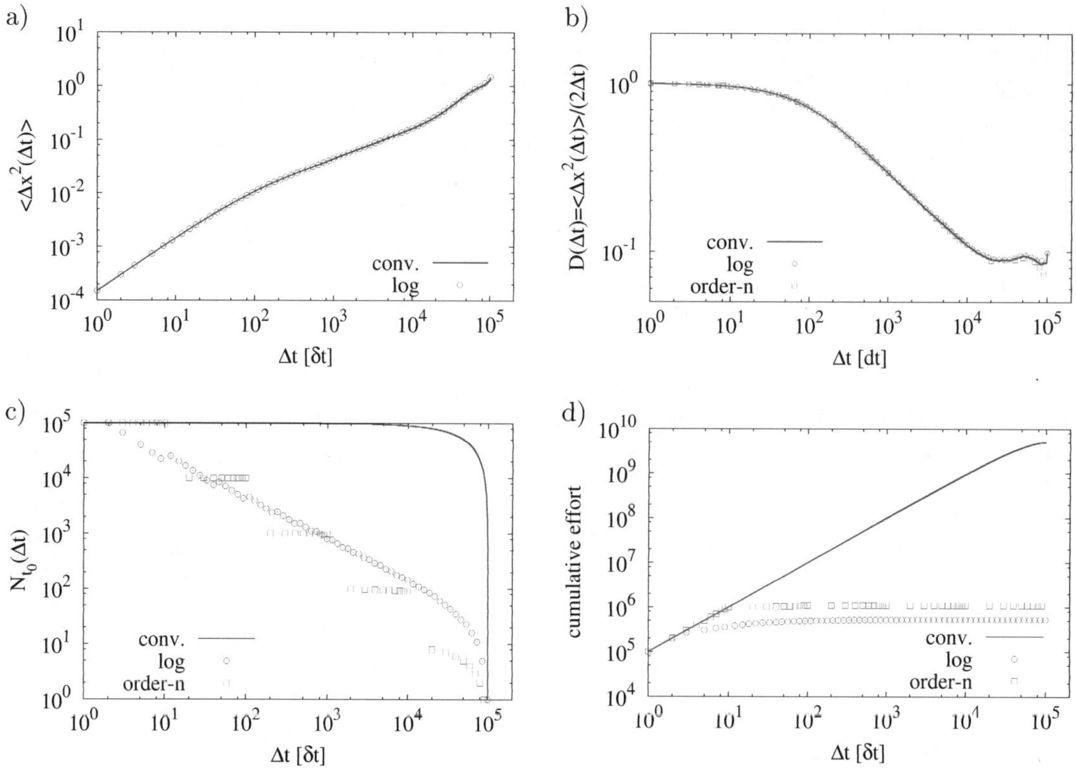

Figure 2. a) The mean square displacement of 300 interacting particles calculated with the log-algorithm (circles) provides comparable accuracy as the MSD calculated with the conventional algorithm (line). b) The time dependent diffusion coefficient shows the deviations of the order-n algorithm (boxes) and the log-algorithm from the results obtained by the conventional algorithm. c) The number of time origins used to calculate the MSD. d) The cumulative effort to calculate the MSD.

all t_0 which are set with different sampling rates R_n.

Figure 2 a) shows the mean square displacement for a simulation of 300 interacting particles with $t_{max} = 10^5$ time steps. The sampling points obtained by the log-algorithm are evenly spaced in the log-log plot, and are close to the curve calculated with the conventional algorithm. In Figure 2 b) a comparison of the time dependent diffusion coefficient $D(\Delta t)$ obtained by the three different methods is given. For the last sampling points the accuracy of the log-algorithm is closer to the conventional algorithm than the order-n algorithm. The reason for this can be seen in Figure 2 c): the log-algorithm uses more time origins than the order-n algorithm for the last few sampling points at large Δt. Since the log-algorithm uses also less time origins for same of the sampling points at small Δt, where the accuracy is already very good, it requires in total less computational time than the order-n algorithm, as can be seen in Figure 2 d).

We provide the algorithm for the MSD calculation within a lattice gas model as test case at https://github.com/ullrichsiems/MSDLOG.

IOP Conf. Series: Journal of Physics: Conf. Series **1012** (2018) 012011 doi:10.1088/1742-6596/1012/1/012011

4. Transport in a channel with a periodic potential

In [20] we studied the transport in a channel with a periodic substrate potential along the channel direction and with a constant pulling force acting on each particles along the same direction. We looked at a channel of width $L_Y = 5.2$ with four layers in the "solid" phase at $\Gamma \approx 25$. The external force acting on the particles is

$$\mathbf{F}_{ext} = F\hat{\mathbf{e}}_x + F_C \cos\left(\frac{2\pi}{\lambda}x\right)\hat{\mathbf{e}}_x\ , \qquad (6)$$

with a maximal reset force $F_C = 41.8$, different period lengths λ, and a pulling force F which was increased from 0 to 100. This system may be pictured as a wave slide for repelling colloidal particles.

The periodic potential has a strength in the same order of magnitude than the interaction

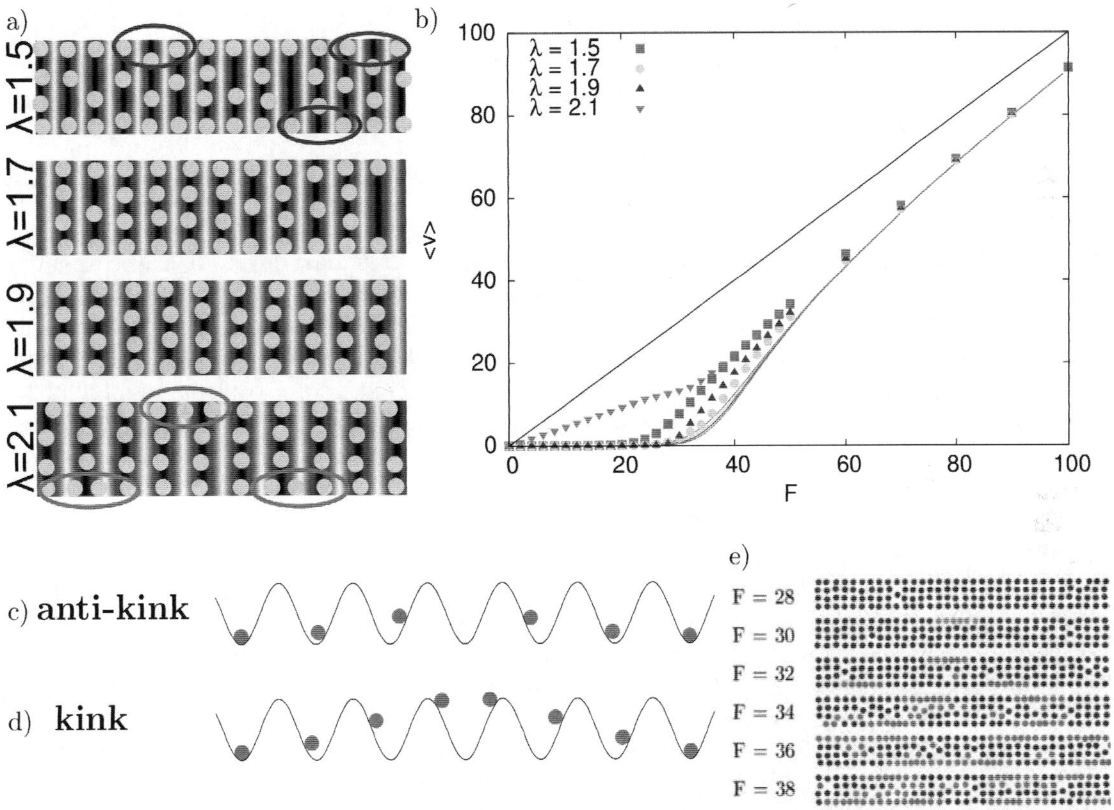

Figure 3. a) Equilibrium particle positions in channels with periodic potentials with different period lengths λ. The potential is color coded with the potential minima in dark. b) Mean particle velocities as a function of the pulling force F for different λ (points) and the solution for a single particle (solid lines). Illustration of c) the kink and d) the antikink soliton of the Frenkel-Kontorova model. e) Snapshot of channel with period length $\lambda = 1.9$ at different pulling forces F, fast particles of moving kinks and anti-kinks are highlighted. (See Ref. [20])

between the particles that enforces the layering parallel to the wall. The potential destroys this layering, and the particles arrange in the minima of the periodic potential (see Fig. 3 a)). Only in the case of $\lambda = 1.9$ the layering with four layers is commensurable with the periodic potential. In this case all minima are occupied by four particles. In the incommensurable cases

there are either vacancies for $\lambda = 1.5$ or additional particles for $\lambda = 2.1$. Both are marked with ellipses (see Fig. 3 a)). The additional particles are located at the boundary layer and they press their neighboring particles a little bit out of the minimum of the periodic potential. This local deformation can be seen as the kink of the Frenkel-Kontorova (FK) model [21] as depicted in Figure 3 d). Whereas the vacancies are the antikinks of the FK model (see Figure 3 c)).
Figure 3 b) shows the mean particle velocities of the particles as a function of the pulling force. In all cases the velocity is increased due to collective effects compared to the solution of a single particle. But in the commensurable case there is no big differences in the mean velocity of the interacting system compared to the single particle picture. For $\lambda = 1.5$ the mean velocity is particularly increased at intermediate forces near the depinning transition, due to the antikinks. The particles can move forward into the vacancy, while the vacancy itself move backwards. For $\lambda = 2.1$ the mean velocity is even more increased. The depinning transition vanishes and the particles already move at low forces, due to the presence of kinks. The extra particles at the boundary can easily be moved, while the attribute of being the extra particles are handed over to the next particle when the particle reaches the next minimum. Therefore the kink excitation moves forwards in the channel faster than the particles themselves. The asymmetry between the kink and antikink can be explained by the purely repulsive interaction. Due to the repulsion kinks can move more easily than antikinks.
How do the particles start to move in the commensurable case with $\lambda = 1.9$, when there is no vacancy a single particle can move into? One could expect that all particles start to move simultaneously at a certain threshold force, but this is not the case. Instead, Figure 3 e) reveals that a particle from one of the mid-layer moves to one of the boundary layer. It leaves an anti-kink on the mid layer and creates a kink on the boundary layer. At the depinning transition, the velocity of the kinks is nearly independent of the external force, but the number of kink and anti-kink pairs increases rapidly. The kink and anti-kinks can pass each other with only a small recombination rate, since they are on different layers. This kind of transport is similar to the conductance in a semiconductor. The kinks at the boundary and anti-kinks on the mid-lane resembles the electrons in the conduction band and holes in the valence band. Although they move in opposite directions, they contribute to a current in the same direction.

5. Diffusion in channels filled with anisotropic particles

As a model for anisotropic particles we used the two-dimensional version of a spherocylinder [22], consisting of a cylinder and two hemispheres rounding off the base and the top of the cylinder. Mathematically a spherocylinder is defined by all points that have the same minimal distance $\sigma/2$ to a line segments of length l, where σ is the diameter of the cylinder base. The aspect ratio of the spherocylinder is $p = (l + \sigma)/\sigma$. The particles interact via the short ranged repulsive Kihara potential [23], which is a function of the minimal distance between the two line segments representing the spherocylinders. The potential has the form of the so called WCA potential which is the repulsive part of the Lennard-Jones potential. The particle-wall interaction is modeled by the same type of interaction.

As an example the results for particles of aspect ratio $p = 2$ are shown. At the beginning of the simulation the particles were organized in four or five layers parallel to the walls, respectively. The width of the channel is varied between $\tilde{L}_Y = 4$ and $\tilde{L}_Y = 5$ at constant particle density. In the early stage of the simulation the particles rearrange according to the width of the channel. When starting with four layers and a width of $\tilde{L}_Y = 4.9$ the particles rearrange in five layers with some skewed particles (see Fig 4 a)). When starting with five layers and a channel width of $\tilde{L}_Y = 4.0$ the spherocylinders rearrange in four layers with some defects (see Fig 4 b)). Even after the rearrangement a orientation along the channel direction dominates. For channels of width $\tilde{L}_Y = 4.5$ the spherocylinders are disordered and orient with a 45 deg angle to the wall. After this quick rearrangement the particles stay in their local cage.

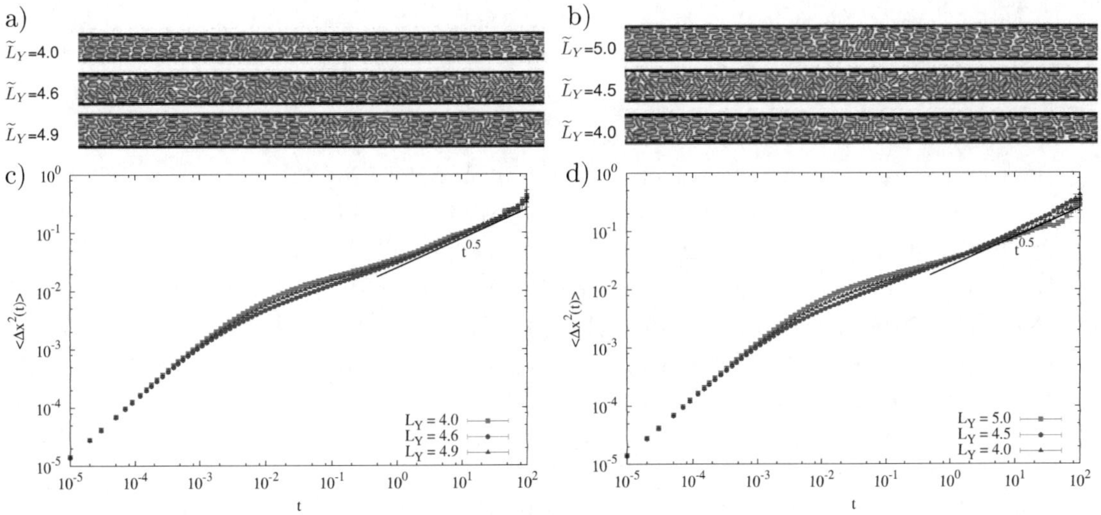

Figure 4. For short spherocylinders with $p = 2$: a) Snapshot after equilibration starting with a configuration of five layers. b) Snapshot after equilibration starting with a configuration of four layers. In c) and d) the mean square displacement along the channel direction is plotted for the above channel systems.

Figure 4 c) and d) show the mean square displacement of the particles along the channel direction after the equilibration phase. In each case the long-time evolution is proportional to $t^{0.5}$, which is a sign of the well known single-file diffusion. Since the system is so dense the particles are locally caged by their neighbors, and only elastic deformations contribute to the mean square displacement. The system can be seen as chain of particles with fixed order, which lead to a single-file diffusion. The intermediate behavior depends on the degree of order and the number of layers in the channel.

Acknowledgments

The authors gratefully acknowledge the SFB 1214 for support and financing the journey to Natal (Brazil), as well as the computing time granted by the John von Neumann Institute for Computing (NIC) and provided on the supercomputer JURECA at Jülich Supercomputing Centre (JSC).

References

[1] Poon W 2004 *Science* **304** 830–831
[2] Bleibel J, Dietrich S, Dominguez A and Oettel M 2011 *Phys. Rev. Lett.* **107** 128302
[3] Helbing D, Farkas I and Vicsek T 2000 *Phys. Rev. Lett.* **84** 1240–1243
[4] Löwen H 1996 *Phys. Rev. E* **53** R29–R32
[5] Zahn K, Méndez-Alcaraz J M and Maret G 1997 *Phys. Rev. Lett.* **79** 175–178
[6] Deutschländer S, Puertas A M, Maret G and Keim P 2014 *Phys. Rev. Lett.* **113** 127801
[7] Illing B, Fritschi S, Kaiser H, Klix C L, Maret G and Keim P 2017 *PNAS* 1856–1861
[8] Heinze B, Siems U and Nielaba P 2015 *Phys. Rev. E* **92** 12323
[9] Squires T and Quake S 2005 *Rev. Mod. Phys.* **77** 977–1026

[10] Yang S Y, Yang J A, Kim E S, Jeon G, Oh E J, Choi K Y, Hahn S K and Kim J K 2010 *ACS Nano* **4** 3817–3822

[11] Ermak D L and McCammon J 1978 *J. Chem. Phys.* **69** 1352–1360

[12] Behringer H and Eichhorn R 2011 *Phys. Rev. E* **83** 065701

[13] Bohlein T, Mikhael J and Bechinger C 2012 *Nat. Mat.* **11** 126–130

[14] Kreuter C, Siems U, Henseler P, Nielaba P, Leiderer P and Erbe A 2012 *J. Phys.: Condens. Matter* **24** 464120

[15] Haghgooie R and Doyle P S 2005 *Phys. Rev. E* **72**(1) 011405

[16] Henseler P, Erbe A, Köppl M, Leiderer P and Nielaba P 2010 *Phys. Rev. E* **81** 041402

[17] Siems U, Kreuter C, Erbe A, Schwierz N, Sengupta S, Leiderer P and Nielaba P 2012 *Scientific Reports* **2** 1015

[18] Frenkel D and Smit B 1996 *Understanding Molecular Simulation* first edition ed (San Diego: Academic Press)

[19] Dubbeldam D, Ford D C, Ellis D E and Snurr R Q 2009 *Molecular Simulation* **35** 1084–1097

[20] Siems U and Nielaba P 2015 *Phys. Rev. E* **91** 022313

[21] Frenkel Y I and Kontorowa T 1938 *Zh. Eksp. Teor. Fiz.* **8** 1340

[22] Löwen H 1994 *Phys. Rev. E* **50** 1232

[23] Kihara T 1953 *Rev. Mod. Phys.* **25** 831–843